SpringerBriefs in Latin American Studies

Series editors

Jorge Rabassa, Ushuaia, Argentina
Eustogio Wanderley Correia Dantas, Fortaleza, Brazil
Andrew Sluyter, Baton Rouge, USA

More information about this series at http://www.springer.com/series/14332

Mario Alberto Hernández
Nilda González · Lisandro Hernández

Hydrogeology of a Large Oil-and-Gas Basin in Central Patagonia

San Jorge Gulf Basin, Argentina

Mario Alberto Hernández
Graduate School of Ecohydrology
National University of La Plata
La Plata
Argentina

Lisandro Hernández
Fundamentals of Geology
National University of La Plata
La Plata
Argentina

Nilda González
Graduate School of Ecohydrology
National University of La Plata
La Plata
Argentina

ISSN 2366-763X   ISSN 2366-7648   (electronic)
SpringerBriefs in Latin American Studies
ISBN 978-3-319-52327-9   ISBN 978-3-319-52328-6   (eBook)
DOI 10.1007/978-3-319-52328-6

Library of Congress Control Number: 2016963173

Printed on acid-free paper

This Springer imprint is published by Springer Nature
The registered company is Springer International Publishing AG
The registered company address is: Gewerbestrasse 11, 6330 Cham, Switzerland

To Nicolás, a little ray of sunshine

Acknowledgements

To Prof. Jorge Rabassa, for his permanent support and sound advice.

To Gabriela Calvetty Ramos, for her valuable graphical work.

To Magdalena Ponce, for her assistance with the English language.

To Rufino Sánchez, Hugo Paoletti, Fernando Perera, Julio Cotti Alegre, Natalia Zanetti and Nazarena Vallines, for contributing invaluable information and comments.

Contents

Abstract

A comprehensive regional study of the groundwater hydrology of the San Jorge Gulf Basin (Argentina) is presented for the first time. It is an onshore area of 40,530 km^2 in an arid region, representative of the Extra-Andean Patagonia, between the Andes and the Atlantic Ocean.

The geohydrology and its basic components—hydrogeology, hydrodynamics, hydrochemistry, and environmental geohydrology—are analysed in a region with precipitations of barely 200 mm/year, a water deficit of 477 mm/year, and a limited occurrence of allochthonous watercourses. Groundwater is therefore essential, especially because it is the most important oil basin in the country, active since 1907, and with excellent prospects regarding the production of unconventional hydrocarbons (shale gas/oil, tight oil), still under development.

A lower or passive geohydrologic system and an upper or active one were recognized. The latter coincides with the most hydrocarbon-productive levels, in sites with ages that span from Early Cretaceous to Pleistocene.

Water in most of the aquifers ranges from brackish to saline, with the Tertiary ones being the only low-salinity units. In order to manage conflicts between uses— in particular between mining and domestic uses—a management and governance plan is proposed for the limited water resources, including low-quality groundwater, the underflow of the Senguerr River, which is the only relatively important watercourse, the reusable water of the injection for secondary recovery and the flow back of the exploitation of unconventional oil and gas (fracking). The use of freshwater aquifers is regulated by provincial protective legislation.

This book also discusses the indispensable groundwater resource protection policies, which are key for the region, and, above all, the need for hydrogeological studies, as at present they are limited or partial.

Keywords Geohydrology · Arid regions · Water deficit · Extra-Andean Patagonia · San Jorge Gulf Basin · Oil/gas exploitation · Unconventional hydrocarbons · Water management and governance

Chapter 1
Introduction

Abstract This first chapter, as well as outlining the purpose of this book, introduces the regional framework of the basin, i.e. the Argentinian Extra-Andean Patagonia—the territory in which the San Jorge Gulf Basin occurs—its geographical location, extension and physical borders. The main geomorphological units are described, together with aspects of the landscape formation and the hydrography. The latter is characterized by its, perennial nature, the nival or pluvio-nival regime of the main rivers, and the fact that they lose water to groundwater, a situation that also occurs in lakes and dams. The lithology of the surface geology is very diverse, with ages ranging from the Proterozoic to the Cenozoic: Tertiary clastic and pyroclastic deposits predominate, followed by Cretaceous sedimentary rocks and Jurassic–Cretaceous igneous rocks. The hydrogeology is mainly conditioned by the climate. There are post-Jurassic aquifers of interest that occur in porous media, as well as deep aquifers occurring in fissured media. Brackish to saline groundwater prevails, with the occurrence of fresh groundwater in the local arid climate being due to particular recharge mechanisms. Its use is restricted and mainly reserved for domestic supply. Hydrocarbon production and metal mining use large volumes, though it is water of lower quality. Such economic activities have displaced the original primary activity, which was sheep farming.

Keywords Purpose · Regional framework · Extra-Andean Patagonia · Regional geomorphology · Geology and hydrogeology · Groundwater quality · Water use · Water-using activities

It is the purpose of this book to characterize from a hydrogeological point of view one of the most special basins in Argentina: subjected to an arid climate, with a complex geology, scarcity of surface and groundwater resources, hosting important hydrocarbon (HC) deposits—both conventional (discovered in 1907) and unconventional—with groundwater historically fulfilling a fundamental role in its socio-economic development.

© The Author(s) 2017
M.A. Hernández et al., *Hydrogeology of a Large Oil-and-Gas Basin in Central Patagonia*, SpringerBriefs in Latin American Studies, DOI 10.1007/978-3-319-52328-6_1

The San Jorge Gulf Basin is undoubtedly a very peculiar geographical and geological feature, located in an even more unique region: the Argentinian Extra-Andean Patagonia. Covering an area of 740,000 km², it is one of the most southern regions on the planet (35°–65° south latitude), with a climate in which the mean annual rainfall values fluctuate between 150 and 300 mm/yr. It is bordered to the north by the Colorado River, to the south by the Strait of Magellan, to the west by the Andes and to the east by the Atlantic Ocean (Fig. 1.1).

It is precisely the characteristics of the boundaries of the regional setting that highlight the above-mentioned singularity, since the western boundary—that is, the Andes—controls the regional climate as it intercepts the Westerlies, discharging up to 2500 mm/yr of rain on the windward side and creating a dry rain shadow to the east (Fig. 1.2).

The Andes, with their permanent snow and continental glaciers, feed the large rivers, nearly all perennial, which traverse the region in search of the ocean (Álvarez et al. 2006), such as the rivers Colorado, Limay-Neuquén-Negro, Chubut, Senguerr, Deseado, Santa Cruz, Coyle and Gallegos (Fig. 1.3).

The eastern boundary is the vast Atlantic Ocean, a discharge area for surface and groundwater runoff; the northern boundary coincides with the limit between the arid Patagonian region and the sub-humid centre; and the southern boundary is the Strait of Magellan, which connects the Atlantic and Pacific oceans.

The regional geomorphology shows a landscape with stepped tablelands descending towards the east from the Andean sector, interrupted by two cratonic massifs, the Somuncurá Massif to the north and the Deseado Massif to the south, and by an orographic belt called "Patagonides", with a trend sub-parallel to the Andes (Coronato et al. 2008). The Patagonides develop from the centre of the region towards the north and they are composed of two sub-units: the Bernardides, in the area of the San Jorge Gulf Basin, and the Patagonian Andean foothills, in the northern sector (Ramos 1999). In Fig. 1.3, such large positive units are shown.

Every process of landscape formation has occurred in this area: *glacial*, mainly in the entire peri-Andean sector; *fluvio-glacial*, in vast extensions of the tablelands covered by gravels attributed to such an origin (Fidalgo and Riggi 1970); *marine*,

Fig. 1.1 Location of the San Jorge Gulf Basin in the Argentinian Extra-Andean Patagonia

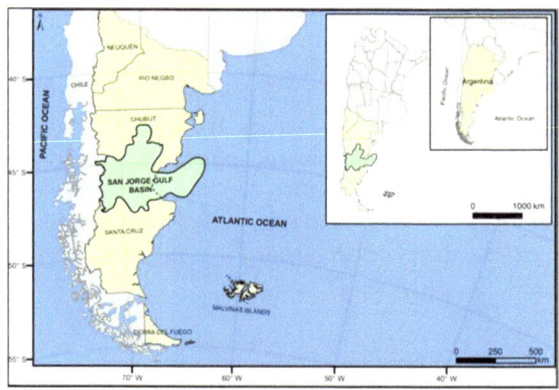

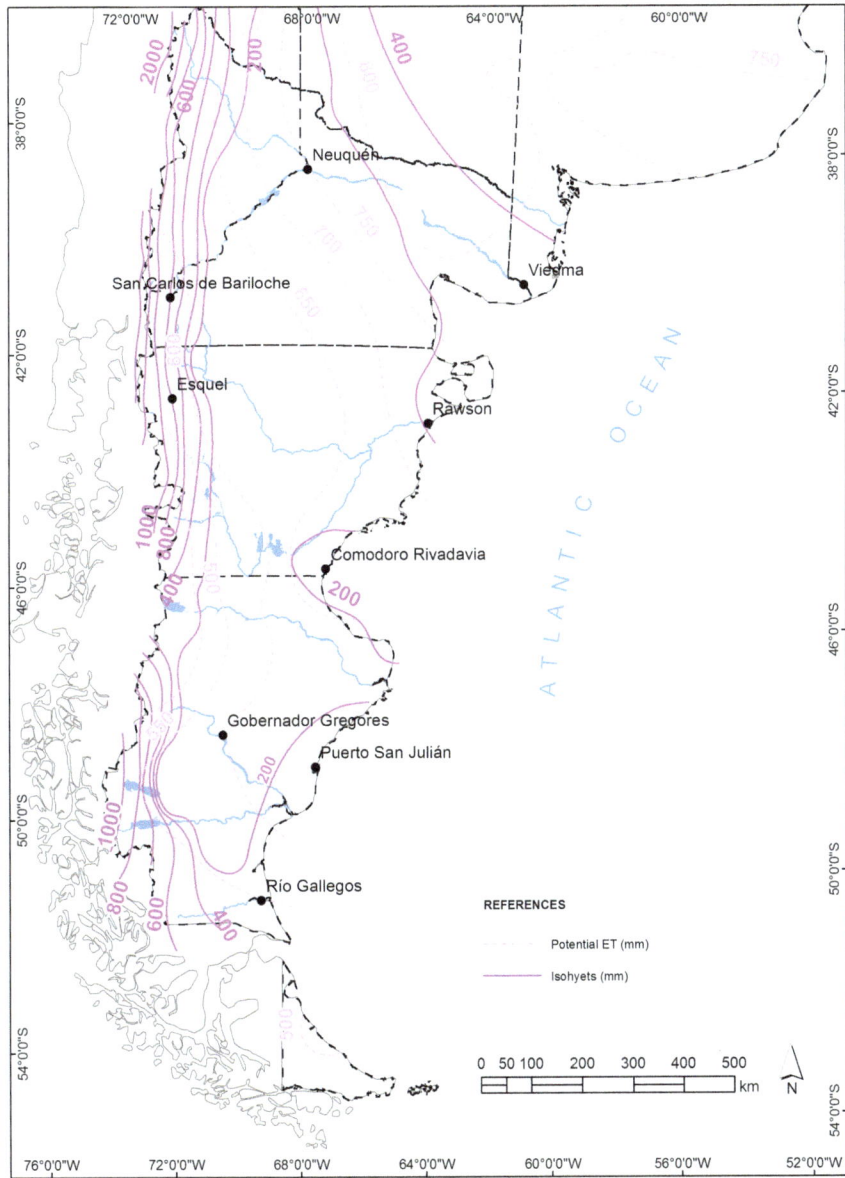

Fig. 1.2 Isohyet and potential evapotranspiration map. The orographic rain shadow of the Andes may be observed, as well as the arid territory to the east, in which the potential evapotranspiration is higher than rainfall

on the eastern boundary; *aeolian*, represented by inland and coastal dunes, some of them active; abundant coastal dunes in plutonic and sedimentary rocks; and *fluvial*, as exhibited in the valleys of the above-mentioned large rivers.

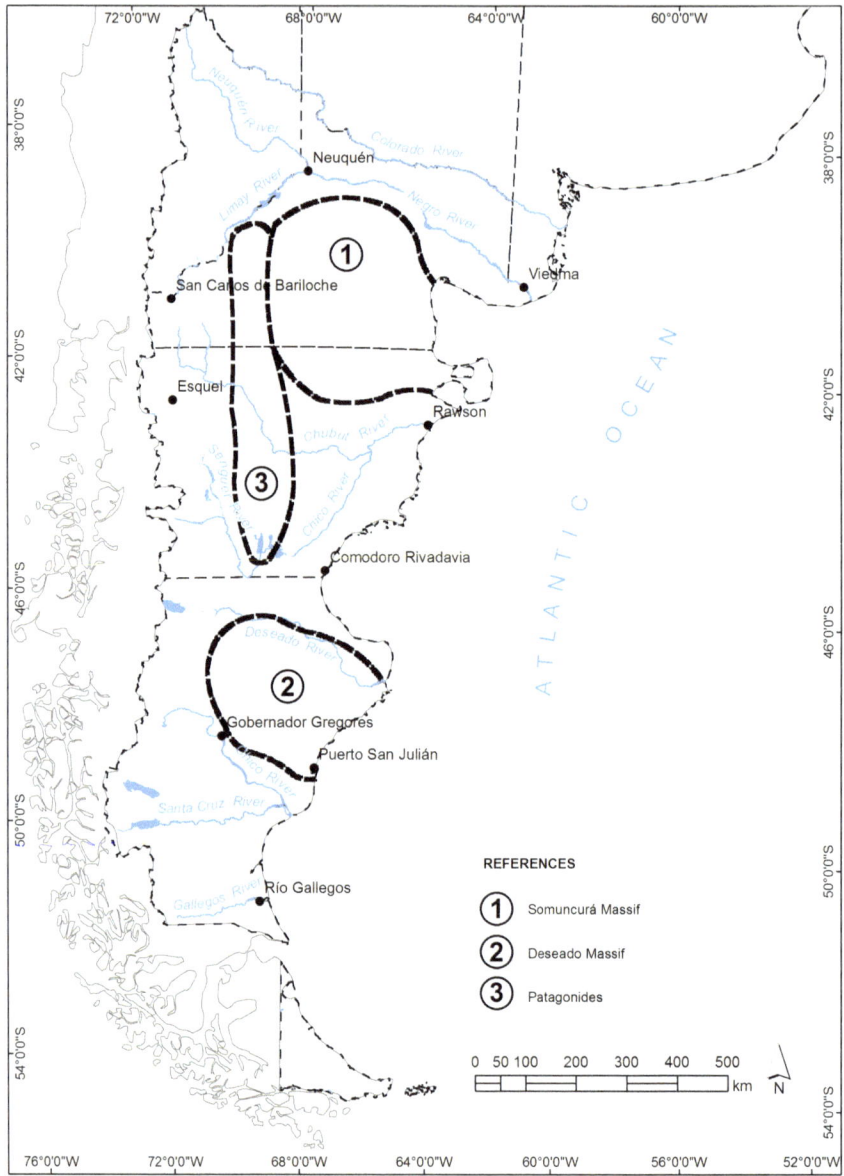

Fig. 1.3 Hydrography and major positive landforms of the Extra-Andean Patagonia

In Fig. 1.3, the main watercourses in the region are also shown: with Andean headwaters, a nival or pluvio-nival regime, perennial for the most part (except for the Chico River), and losing or influent—as they lose water to groundwater—with a discharge that reaches 1014 m³/s in the case of the Negro River. The most

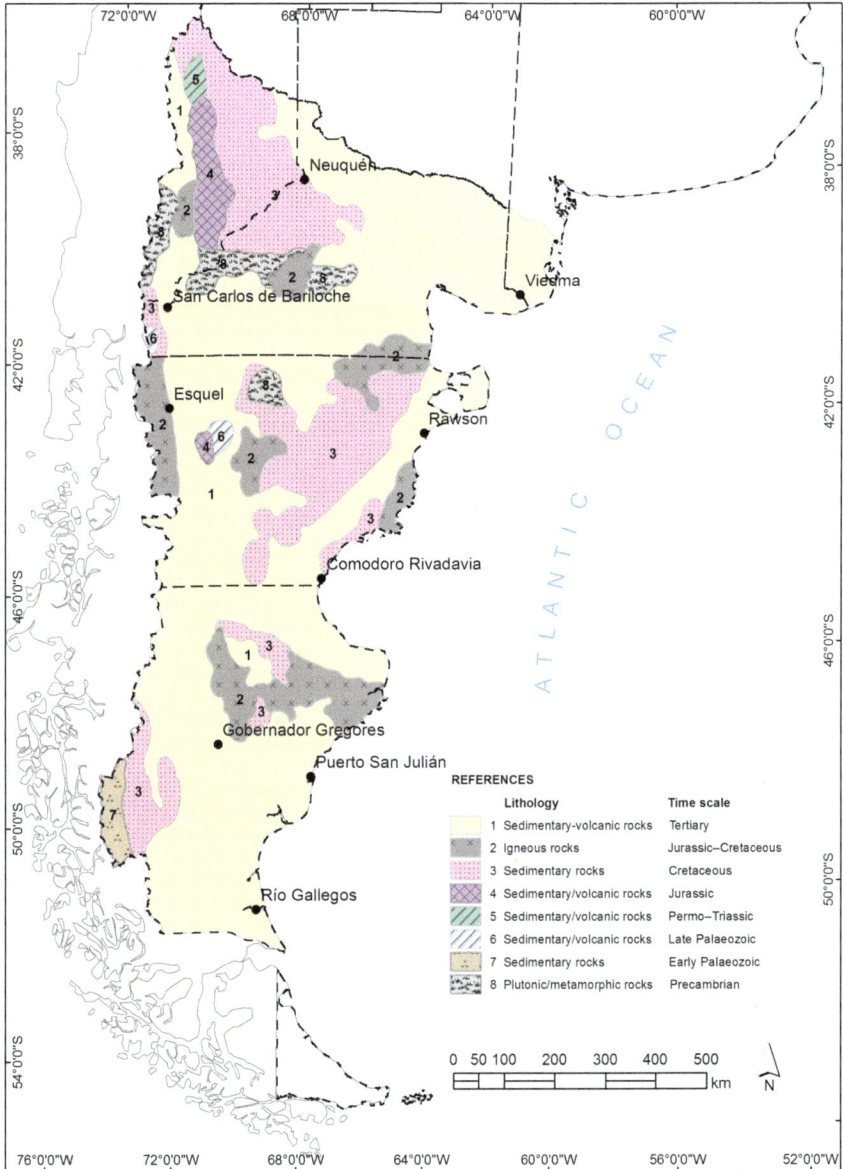

Fig. 1.4 Surface geological sketch map, showing the main lithological types and their age assignation (time scale)

important extra-Andean waterbodies are the Musters, Colhué Huapi (both of these in the San Jorge Gulf Basin), Cardiel, and Strobel lakes, as well as the Alicurá, Piedra del Águila, El Chocón, Mari Menuco and Los Barreales reservoirs.

The outcropping geology includes the most diverse lithologies and ages ranging from the Proterozoic to the Cenozoic. Figure 1.4 shows a sketch map representing the most important lithological units according to their age.

It is possible to observe that, over most of the region, Tertiary[1] clastic (gravels and sands) and pyroclastic deposits crop out, followed in abundance by Cretaceous sedimentary rocks and Jurassic–Cretaceous igneous rocks. The former units prevail in the basin, as will be shown below, together with the detailed subsurface geology, which is also representative of the geological time scale and the different lithological types.

From a hydrogeological point of view, the above-mentioned climate is a fundamental determining factor of groundwater quality and quantity. The post-Jurassic levels are of special interest: they are characterized by a medium with primary porosity, whereas the rest of the geological column includes aquifers in a fissured medium. Recharge is regulated by particular mechanisms that explain the occurrence of fresh groundwater in an arid climate in the Extra-Andean Patagonia (Hernández et al. 2008, 2009).

In keeping with the limited availability, groundwater use is restricted and mainly intended for domestic supply. On the other hand, HC production and metal mining use large volumes, though of inferior quality.

The economy, originally based on sheep farming, changed radically with the discovery of oil in the early twentieth century and the development of metal mining (mainly gold mining) in the last decades of the last century, whereas the price of wool fell and therefore sheep production declined. At present, oil and gas production is the main socio-economic activity, followed by gold mining, intensive agriculture, fishing, industry and livestock farming.

In an intermediate latitude in this region is the San Jorge Gulf Basin (Fig. 1.1), whose essential features relevant to the subject of this book shall be described, taking into consideration the physical, hydrological and socio-economic contexts that have been briefly described.

References

Álvarez MP, Hernández L, Hernández MA, González N (2006) Relación aguas subterráneas-aguas superficiales en Patagonia Extrandina. República Argentina. Revista Latino-Americana de Hidrogeología 6:43–48 (Montevideo, Uruguay)

Coronato A, Coronato F, Mazzoni E, Vázquez M (2008) The physical geography of Patagonia and Tierra del Fuego. In: Rabassa J (ed) The Late Cenozoic of Patagonia and Tierra del Fuego, Developments in Quaternary Science, vol 11. Elsevier, Amsterdam, pp 13–55

---

[1]The authors find it necessary to clarify that in the text it was decided to use the informal term "Tertiary" instead of the formal ones "Palaeogene" and "Neogene" in order to retain the original term used by the cited authors (some of them in recent contributions) and to preserve the quotations and other references, such as basic charts and tables.

Fidalgo F, Riggi JC (1970) Consideraciones geomórficas y sedimentológicas sobre los Rodados Patagónicos. Revista Asociación Geológica Argentina 25(4):430–443 (Buenos Aires)

Hernández L, Hernández MA, González N, Ceci JH, Sánchez R (2008) Origen de aguas subterráneas salinas en la zona de Caleta Olivia. Provincia de Santa Cruz, Argentina. IX Congreso Latinoamericano de Hidrología Subterránea, ALHSUD. Quito [CD-ROM]

Hernández MA, González N, Hernández L (2009) Regiones áridas. Procesos diferenciales de recarga y casos ejemplo de Argentina. In: Carrica J, Hernández MA, Mariño E (eds) Recarga de acuíferos. Aspectos generales y particularidades en regiones áridas. AIHGA-Amerindia, Santa Rosa, pp 63–70

Ramos V (1999) Las provincias geológicas del territorio argentino. In: SEGEMAR (ed) Geología Argentina, Anales. Instituto de Geología y Recursos Minerales, Buenos Aires, vol 29, no 3, pp 41–96

Chapter 2
Methodology

Abstract An overview of the methodology is presented; according to the deductive, plausible reasoning, the results are the most plausible hypotheses, not categorical conclusions. The criterion of convergence of evidence from different fields of knowledge was adopted. Concerning the hydrometeorology, the main meteorological phenomena were analyzed, calculating evapotranspiration (potential and actual), as an approach to the water balance. The Comodoro Rivadavia weather station (1921–2010) was used, as it has the reliability, continuity and representativeness recommended by the WMO. Geomorphology, surface geology and soil characterizations were based on the data collected personally or by several authors. Subsurface geology was reconstructed using geological and geophysical well logs. Lithostratigraphic units were translated into hydrolithological units to define the physical component of the geohydrological system. Regarding the mobile component, recharge, circulation, and discharge were analyzed. As to the first process, the net vertical input was estimated, based on a serial water balance, classified rainfall events, and the specific phenomena occurring in arid areas. Circulation and discharge were identified by the potentiometric surface of wells and the construction of an equipotential diagram. The hydrochemical data was processed using the Easy_Quim 5.0 software to convert units to milliequivalents, calculate analysis error, construct specific plots and calculate ion ratios. Having defined the conceptual model and estimated the availability, the socio-economic context was analyzed. Attention was given to oil and gas production, including unconventional HC exploitation and the participation of groundwater, with a focus on conflicts between uses and how to manage them in the context of water governance.

Keywords Deductive methodology · Plausible reasoning · Geohydrological system · Arid regions · Procedures

Conceptually, the general methodology used is deductive in character and, given that the fundamental subject of this research—that is, water—is eminently dynamic, the approach adopted is that of plausibility. The results obtained will therefore be the most plausible hypotheses, and not definitive and unchanging conclusions; in

© The Author(s) 2017
M.A. Hernández et al., *Hydrogeology of a Large Oil-and-Gas Basin in Central Patagonia*, SpringerBriefs in Latin American Studies, DOI 10.1007/978-3-319-52328-6_2

order to do so, the criterion of convergence of evidence from different fields of knowledge had to be applied.

As regards the subject of hydrometeorology, the analysis of the main meteorological phenomena (rainfall, temperature, wind, barometric pressure, relative humidity, snowfall and frost) was carried out, as well as the calculation of the evapotranspiration (potential and actual), with the consequent approximation to the water balance. At that point, the first difficulties arose, due to the great scarcity of basic information, since out of six stations of the Servicio Meteorológico Nacional (SMN; National Weather Service) that keep records within the basin, only two (Comodoro Rivadavia and Sarmiento) meet the World Meteorological Organization (WMO) requirements, whereas the records in the others are discontinuous, incomplete, or too brief.

Out of the nearby weather stations, the only ones that could have been used (Trelew, Puerto Madryn, Puerto Deseado and Gobernador Gregores) are distant or do not comply with the standards. However, they were useful for control and as a complement, but not for their application in equations. That leaves only Comodoro Rivadavia as usable, with records between 1921 and 2010. This weather station meets the conditions of *continuity* (90 years, as opposed to the 30 years recommended by the WMO), *reliability* (the data are cleaned by the SMN), and *representativeness* (as it is located in the geographical centre of the Gulf). Modular rainfall graphs were generated for the period and for decadal variations. For the other meteorological phenomena of interest, some illustrative graphs were also developed and, in the case of winds, graphs of velocity and direction.

One product of the hydrometeorological analysis is the water balance, undertaken by means of the well-known method by Thorthwaite and Mather (1955), which, even though it is not recommended for regions with extreme climates, at least facilitates the identification of periods with a possibly lower water deficit. In turn, the estimation of probable autochthonous contributions to the groundwater system could be carried out by means of the comparison with serial water balances using the Balshort software (Carrica 1993). It was applied to Puerto Madryn, which is near the basin, for a shorter period and, as it is a daily time-step model, it offers the possibility of quantifying the occasional winter excesses. It was also implemented in other nearby cases, using classified rainfall events (Hernández 2000); these are analyzed, together with the above-mentioned case, in Sect. 3.1.

The characterization of the geomorphology, surface geology, and soils was undertaken on the basis of personally collected information and other data produced by several authors, such as Feruglio (1949, 1950), Lesta et al. (1980), Cesari et al. (1986), Consejo Federal de Inversiones (1986), Homovc and Lucero (2002), Coronato et al. (2008), Hernández et al. (2008, 2009), Hernández et al. (2008), and Sylwan et al. (2011), among others. The personally collected data for different sectors of the basin, especially in the oil-bearing area, were corroborated and organized by region by means of freely available satellite imagery.

Subsurface geological information was more abundant, specifically in the oil and gas exploration and exploitation area (between the Deseado River, the Senguerr-Chico fluvial system and the 70° W meridian). Even though very good data could be obtained from the geological profiles compiled, and from their interpretation and correlation, there is a confidentiality issue that limits their use, especially at depth. The contributions collected in Schiuma et al. (2002) and Sylwan et al. (2011) were of particular significance, as they were very helpful as regard this topic.

The main focus was on the lithological descriptions of the geological and geophysical profiles; the latter were used to identify transitions between the different units. Within the scope of geohydrology, the former were useful to transform the lithostratigraphic units (e.g. formations, members, etc.) into lithostratigraphic units (i.e. aquifers, aquicludes, aquitards and aquifuges).

The physical component of the geohydrological system of the basin could thus be identified and defined, verifying the conformation described by Hernández and Hernández (2013): an overlying basement constituted by the Jurassic Volcano-Sedimentary Complex (Complejo Volcánico-Sedimentario; Sylwan et al. 2011), with Precambrian and early Mesozoic igneous sections and pre-Jurassic sedimentary rocks, depending on the region under consideration.

As regards the mobile component of the system, and based on a conceptual model, the processes of recharge, circulation and discharge were analyzed. In the case of recharge, and given the extremely arid climate of the region, in order to estimate the net vertical input, extrapolations from a serial water balance and classified rainfall events were used, as mentioned above. But, above all, the occurrence of specific phenomena described by Hernández (2015) and Hernández et al. (2008, 2009) was recognized, such as a reduction in consumptive losses, rapid infiltration, rapid concentration, fluvial influence and delayed recharge.

The groundwater circulation and discharge phenomena were identified by using the potentiometric surface, constructed with values of the water level in wells generally used for HC exploitation, for water injection in oil fields, or drilled for environmental purposes by the oil industry; this information is therefore concentrated in the central and eastern sectors of the basin. Such values were calculated as the difference between the topographic height at well head and the depth measured in each, expressed with respect to the sea level.

The main difficulty was caused, apart from the spatial distribution of the information, by the characteristics of the wells regarding the validity of the hydrometric data, which led to a validation process in order to discard those with uncertainty. It should be taken into consideration that these wells are of interest to the oil industry, and that often there is no record of the aquifer sections of the wells of exclusive hydrological interest.

The result was an equipotential diagram based on the heights, with an equidistance of 100 m, given the scale and size of the area, and indicating flow directions. Another inconvenient was the lack of data on permeability (K) and transmissivity (T) coefficients, due to the absence of hydraulic tests adequate for quantitative purposes. In any case, they were sufficient to define the processes, estimate the flow velocity as a function of K, and establish the geohydrological conceptual model.

The same applies to the hydrochemistry, since—with few exceptions—the analytical information derived from the oil industry. In order to process such data, the Easy_Quim 5.0 (Vazquez-Suñé and Serrano-Juan 2012) free software was used. It is applied to the conversion of content units in terms of weight to milliequivalents (mg/L to meq/L), the calculation of analysis error and the validation of the data. It also allows the plotting of graphs with the results by means of the Piper, Schoeller-Berkaloff, and Stiff methods (Custodio and Llamas 2001). Besides, it offers ion ratios of interest, such as rNa^+/rK^+, rMg^{++}/rCa^{++}, $rSO_4^=/rCl^-$, rCl^-/rCO_3H^- and the ionic exchange index.

Once the conceptual model had been defined and the availability had been estimated, the socio-economic context was analyzed, with a focus on the production of oil and gas, which was characterized from its beginning in the study area.

The analysis included the involvement of groundwater in the different stages of oil projects: exploration, production, transportation, transformation and marketing, both regarding groundwater use and the impact of this activity on the environment and on aquifer protection. It aimed at focusing on the actual and potential conflicts between uses, and on how to attempt to manage them in the context of water governance.

A special circumstance that arose in the last few years relates to the real possibilities for the production of unconventional oil and gas from well-known source rocks. This has called for renewed efforts to obtain groundwater, given the larger volumes required and the above-mentioned scarcity of the resource. Therefore, a new methodological approach is developed, aiming at management optimization with the introduction of new tools.

References

Carrica JC (1993) El Balshort. Un programa de balance hidrológico diario del suelo aplicado a la región sudoccidental pampeana. XII Congreso Geológico Argentino, Actas VI:243–248 (Mendoza)

Cesari O, Simeoni A, Beros C (1986) Geomorfología del Sur de Chubut y Norte de Santa Cruz. Revista Universidad Nacional de la Patagonia. Comodoro Rivadavia, vol I, no 1, pp 18–36

Consejo Federal de Inversiones (1986) Geología y Geomorfología del NE de la Provincia de Santa Cruz. Consejo Federal de Inversiones Tomos I y II, Buenos Aires 89 pp

Coronato A, Coronato F, Mazzoni E, Vázquez M (2008) The Physical Geography of Patagonia and Tierra del Fuego. In: Rabassa J (ed) The Late Cenozoic of Patagonia and Tierra del Fuego, Developments in Quaternary Science. Elsevier, Amsterdam, vol 11, pp 13–55

Custodio E, Llamas MR (2001) Hidrología Subterránea, Tomo I and II, 2nd edn. Omega, Barcelona 2350 pp

Feruglio E (1949) Descripción geológica de la Patagonia. Dirección General de Yacimientos Petrolíferos Fiscales, vol 1, no 2. Buenos Aires

Feruglio E (1950) Descripción geológica de la Patagonia. Dirección General de Yacimientos Petrolíferos Fiscales, vol 3, Buenos Aires

Hernández MA (2000) Estudio geohidrológico de la región Cerro Rubio-Cerro Vanguardia, provincia de Santa Cruz. Unpublished doctoral thesis, Facultad de Ciencias Naturales y Museo, Universidad Nacional de La Plata, 163 pp

Hernández MA (2015) Recursos hídricos. In: Recursos hidrocarburíferos No Convencionales shale y el desarrollo energético de la Argentina. Caracterización, oportunidades, desafío. EUDEBA, Buenos Aires, vol 4, pp 307–348

Hernández L, Hernández MA (2013) Características hidrolitológicas de las formaciones Patagonia y Santa Cruz. Cuenca del Golfo San Jorge. (Provincias de Chubut y Santa Cruz). In: González N, Kruse EE, Trovatto MM, Laurencena P (eds) Agua subterránea recurso estratégico. EDULP, La Plata, vol I, pp 112–117

Hernández MA, González N, Hernández L (2008) Late Cenozoic geohydrology of Extra-Andean Patagonia Argentina. In: Rabassa J (ed) The Late Cenozoic of Patagonia and Tierra del Fuego, Developments in Quaternary Science. Elsevier, Amsterdam, vol 11, pp 497–509

Hernández MA, González N, Hernández L (2009) Regiones áridas. Procesos diferenciales de recarga y casos ejemplo de Argentina. In: Carrica J, Hernández MA, Mariño E (eds) Recarga de acuíferos. Aspectos generales y particularidades en regiones áridas. AIHGA-Amerindia, Santa Rosa, pp 63–70

Homovc JF, Lucero M (2002) Cuenca del Golfo San Jorge: Marco geológico y reseña histórica de la actividad petrolera. In: Rocas reservorio de las cuencas productivas de la Argentina. Instituto Argentino de Petróleo y Gas (IAPG), Buenos Aires, pp 119–126

Lesta PJ, Ferello R, Chebli G (1980) Chubut extrandino. In: Turner JCM (coordinator): Segundo Simposio de Geología Regional Argentina. Academia Nacional de Ciencias de Córdoba 2:1307–1387

Schiuma, M, Hinterwimmer, G, Vergani G (eds) (2002) Rocas reservorio de las cuencas productivas de la Argentina. V Congreso de Exploración y Desarrollo de Hidrocarburos. Instituto Argentino de Petróleo y Gas (IAPG), Mar del Plata

Sylwan C, Droeven C, Iñigo J, Mussel F, Padva D (2011) Cuenca del Golfo San Jorge. VIII Congreso de Exploración y Desarrollo de Hidrocarburos. Simposio Cuencas Argentinas: visión actual, pp 139–183 Instituto Argentino de Petróleo y Gas (IAPG). Mar del Plata

Thornthwaite CW, Mather JR (1955) The water balance. Drexel Institute of Technology, Laboratory of Technology, Publications in climatology, Centerton, NJ, vol 8, no 1, 104 pp

Vazquez-Suñé E, Serrano-Juan A (2012) EASY_QUIM v. 5.0 www.h2ogeo.upc.edu

Chapter 3
San Jorge Gulf Basin

Abstract The basin is first characterized by its hydrometeorology: using as reference the Comodoro Rivadavia Aero weather station, the main variables are analyzed statistically and graphically. The mean annual precipitation is 227 mm, with a decadal upward trend, and the mean annual temperature reaches 12.7 °C. Westerly winds prevail (32 km/h), the relative humidity is 49%, and the mean barometric pressure is 1000.2–1006.0 hPa. Potential evapotranspiration reaches 7004 mm/yr, with an annual deficit of 477 mm. The climate is arid, mesothermal with no water excess. The basin is a part of the South American Plate and the regional sedimentary environment developed during the late Carboniferous–Permian. In the Jurassic, an extensional tectonic event occurred, filled with Cretaceous and Palaeogene sediments in a late-rift stage. The major positive landforms are the Patagonian Tablelands, Bernardides, low hills, littoral ridges and dunes. The negative ones, the depression of the central lakes, the valleys of the Senguerr, Chico, Deseado, and Chubut rivers and endorheic depressions. The Senguerr is the only intra-basin perennial watercourse. Aridisols dominate over Entisols and Mollisols, with xerophytic vegetation. The deepest rocks are metamorphic and Palaeozoic–Precambrian intrusives; Devonian granites and schists; Carboniferous–Permian sedimentary rocks; Permian–Triassic igneous rocks; Triassic pelites, psammites and pyroclastites; and Jurassic Volcano-Sedimentary Complex rocks. Then, Lower Cretaceous and Chubut Group deposits occur, hosting the main oil-bearing formations. A Palaeogene–Neogene sequence follows, topped by the Patagonia and Santa Cruz formations, where brackish or freshwater aquifers occur. The column ends with Cenozoic sediments and/or modern basalt layers.

Keywords San Jorge Gulf Basin · Hydrometeorology · Geological history · Geomorphology · Soils · Hydrography · Geology

First of all, it should be clarified that the term "basin" is more of a geological than a physiographic concept, since, for example, it extends several miles offshore and it is even referred to as a "hydrocarbon-producing basin". Besides, the text basically refers to the extra-Andean area, with the Andes being regarded as a boundary.

© The Author(s) 2017   15
M.A. Hernández et al., *Hydrogeology of a Large Oil-and-Gas Basin in Central Patagonia*, SpringerBriefs in Latin American Studies, DOI 10.1007/978-3-319-52328-6_3

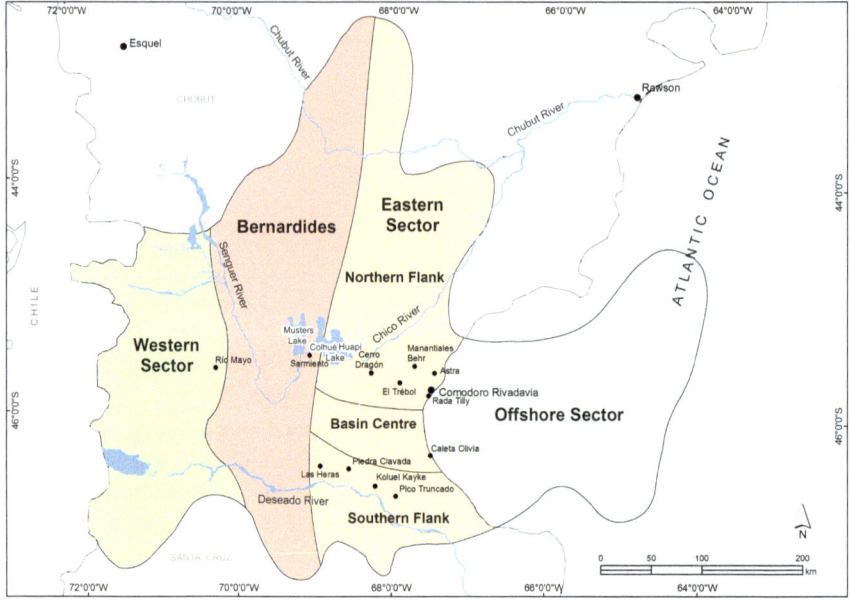

Fig. 3.1 Map of the San Jorge Gulf Basin, indicating sectors, subsectors, and main localities

It extends over an area of approximately 59,510 km^2, out of which 40,530 km^2 are onshore and the rest, offshore. The territory is divided between the provinces of Chubut and Santa Cruz, with its socio-economic centre being the city of Comodoro Rivadavia (186,583 inhabitants) and including important localities, such as Caleta Olivia (51,733), Las Heras (23,604), Pico Truncado (20,889), Sarmiento (10,858), Rada Tilly (9100), Perito Moreno (4617), Los Antiguos (3963), Alto Río Senguerr (1693) and other smaller settlements, adding up to 310,000 inhabitants in the year 2010, in the last national census (INDEC 2011), and an estimated 380,000 in 2015. In Fig. 3.1, the location of the above-mentioned localities is shown.

As a general characterization, the hydrometeorological, geomorphological, surface hydrological, edaphological, and geological aspects are described below, preceding the geohydrological discussion, which is the main subject of this book.

3.1 Hydrometeorology and Climate

As anticipated in the introduction, the climatic factor is the key determining factor in the geohydrology of the region. In this chapter, the main meteorological phenomena that define it are analyzed, using data from the SMN's Comodoro Rivadavia Aero weather station (45°47′ S, 67°30′ W; with a height of 61 m a.s.l.), which has a 90-year record (1921–2010). In order to compare behaviours, a coastal station (Puerto Madryn) and another one located entirely inland (Gobernador Gregores) were used; both of them are within the Extra-Andean Patagonian region. Precipitation, rainfall, temperature, wind (velocity, frequency, and direction), relative humidity, and snowfall were analyzed due to their specificity. In all cases, the data used were provided by the SMN.

The mean annual precipitation is 227 mm, concentrated in the autumn–winter cold semester (147 mm and 64%), with a mean maximum in May (34 mm and 15%) and minimum in October and December (12 mm and 5.3%). In Fig. 3.2, a bar chart is shown, clearly reflecting the Pacific rainfall regime in a unimodal hyetograph. This winter concentration of rainfall is significant, which will be clear below when compared with the behaviour of other meteorological phenomena (wind, relative humidity, and barometric pressure).

The variation in the decadal average rainfall for the periods available (Fig. 3.3) shows an upward trend, with one exceptional decade (1971–1980).

Figure 3.4 shows the temperature graph (mean annual temperature) for the period 1921–2010. The average modular temperature reaches 12.7 °C, with a maximum in January (18.9 °C) and a minimum in July (6.7 °C).

It is interesting to observe in Fig. 3.5 that the temporal evolution of the decadal mean temperatures shows a similar trend to the one of rainfall, although more marked.

Winds have an influence, not only because they condition the regional climate, as mentioned above, but also due to their local impact in the dissipation of water vapour. Their mean velocity and quadrant frequency are of interest: the former is high, with mean annual values of the order of 32 km/h, average maximum values of 35–39 km/h in the summer months (November to January) and winter average

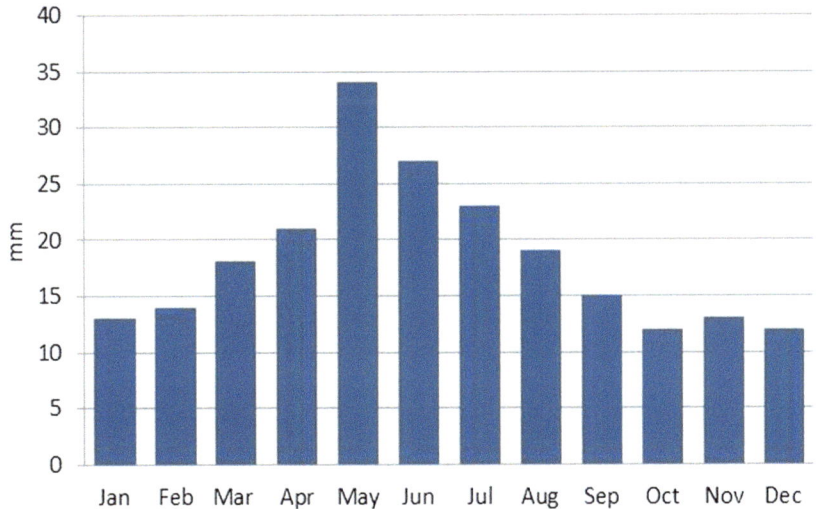

Fig. 3.2 Modular hyetograph for 1921–2010. Comodoro Rivadavia Aero Station (SMN)

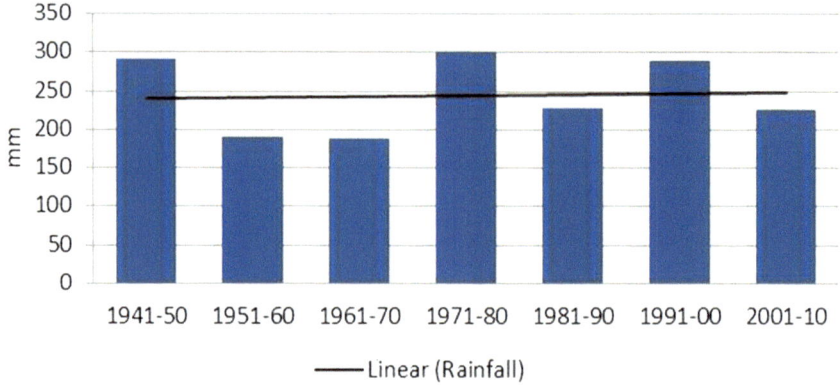

Fig. 3.3 Variation in decadal average rainfall. Comodoro Rivadavia Aero Station (SMN)

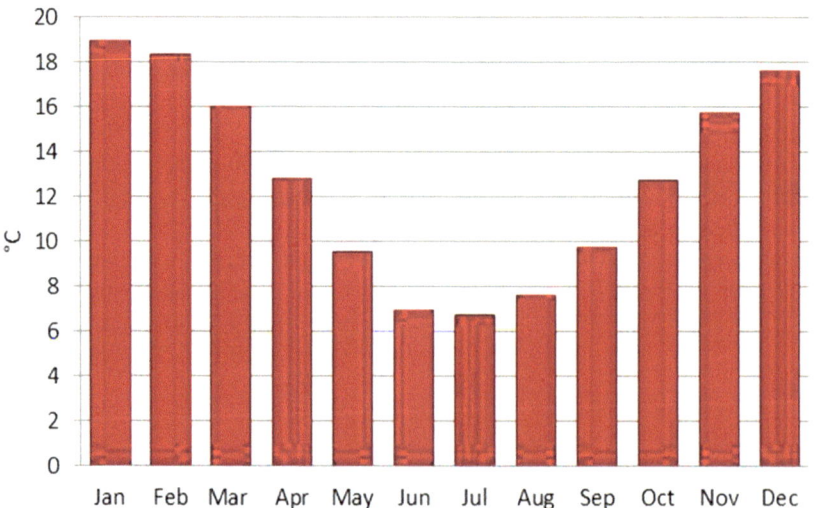

Fig. 3.4 Modular temperature graph for 1921–2010. Comodoro Rivadavia Aero Station (SMN)

minimum values of 28–32 km/h (July). The prevailing winds are from the west, with an annual frequency of 473/1000, followed by those from the southwest (111/1000) and northwest (97/1000), the latter only barely exceeding the calm conditions (100/1000). The influence of the Westerlies is clear, since winds from the west quadrant predominate (Fig. 3.6).

The annual relative humidity value is 49%, with maximum values occurring in winter: over 50% between March and September and a peak in July (59%). Minimum values occur in the summer, with the lowest corresponding to December and January (41%).

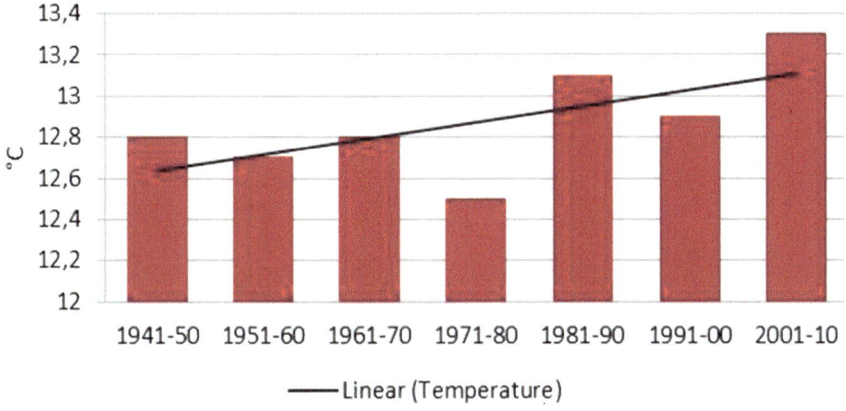

Fig. 3.5 Variation in decadal mean temperatures. Comodoro Rivadavia Aero Station (SMN)

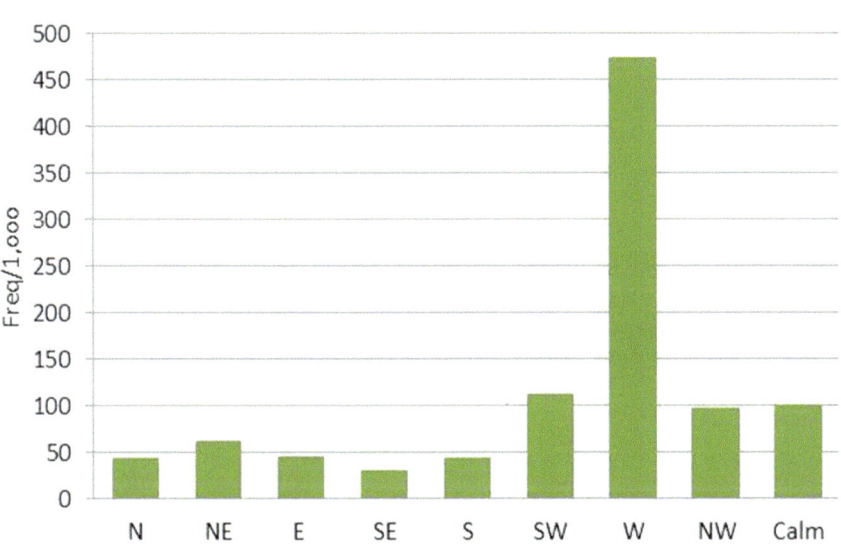

Fig. 3.6 Mean wind frequency distribution by quadrant for 1941–2010. Comodoro Rivadavia Aero Station (SMN)

The same applies to barometric pressure, although with a lower range of variation between the extremes of 1000.2 and 1006.0 hPa.

This coincidence of higher precipitation, relative humidity and barometric pressure in the cold season, with minimum wind velocity values (calm conditions), had already been anticipated for a region towards the south of the basin (Cerro Rubio-Cerro Vanguardia) by Hernández (2000). It is regarded as a favourable factor for infiltration in the Extra-Andean Patagonia, within the context of the recharge mechanisms that are described in Chap. 4 (Fig. 3.7).

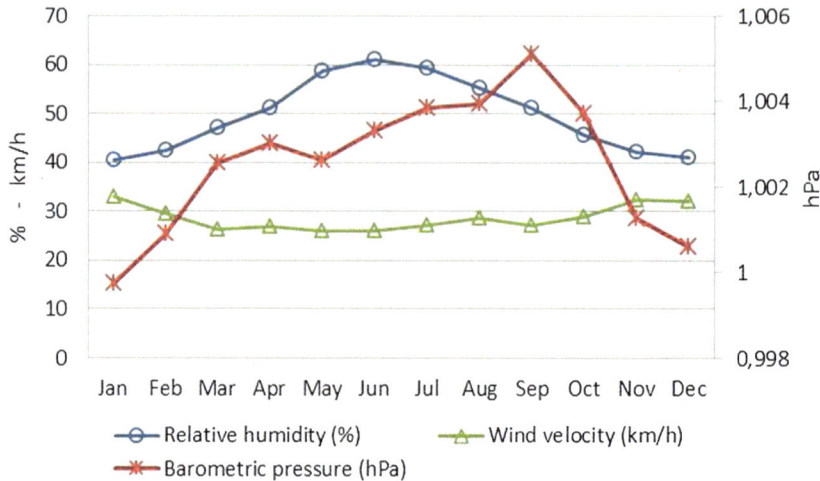

Fig. 3.7 Variations in mean relative humidity, mean wind velocity, and mean barometric pressure for 1941–2010. Comodoro Rivadavia Aero Station (SMN)

In order to estimate potential evapotranspiration and actual evapotranspiration, the Thornthwaite and Mather (1955) method was used. These authors themselves recognize that it should not be applied as formulated if it is intended to calculate a water balance in extreme climates, such as the arid climate of the San Jorge Gulf Basin. Therefore, as clarified in Chap. 2, the objective was to identify the months in which the lowest water deficits occurred, as well as using such data in climate classification. Useful water reserves in the soil of 100 and 50 mm were used.

In Fig. 3.8, the results of this method for the SMN's Comodoro Rivadavia Aero Station have been plotted for the period 1921–2010. Annual potential evapotranspiration reaches 704 mm, concentrated in the warm semester October–March (532 mm and 75.6%), with a peak in January (115 mm). The minimum monthly value (17 mm) occurs in June.

In observing the figure, the large difference between precipitation and potential evapotranspiration can be noticed, that is, an annual deficit of 477 mm, occurring between August and April. The only months with no deficit are May, June, and July, and in August, it barely reaches 2 mm.

It can be appreciated that actual evapotranspiration logically has the same value as precipitation, as there are no surpluses in the balance. Therefore, it can be seen that the Thornthwaite and Mather (1955) formula does not apply when quantifying the contributions to the aquifer system, a subject that will be taken up in Chap. 4, when specifically discussing the phenomenon of recharge.

The same methodology can be used for climate classification (Burgos and Vidal 1951), using the aridity index (100 · water deficiency/water need = 67.76%), humidity index (100 · water surplus/water need = 0), water index (100 · water

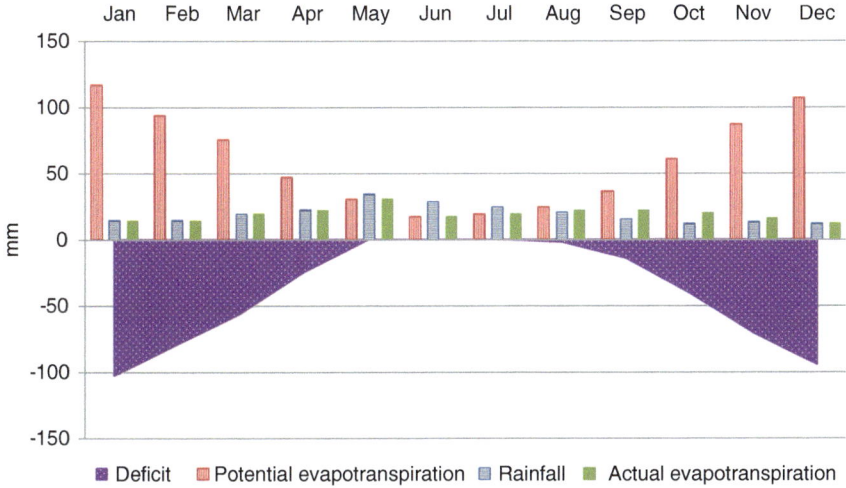

Fig. 3.8 Modular water balance for 1921–2010. Comodoro Rivadavia Aero Station (SMN)

surplus − 60 · water deficiency/water need = −40.65), and the summer concentration of thermal efficiency = 43.46%. The type of climate is therefore *arid, mesothermal with no water excess and a summer concentration of thermal efficiency of less than 48%* (E B′₁da′).

3.2 Geological History

The basin is within the geotectonic domain of the South American Plate, at its southern end. The regional sedimentary event that originally included the basin corresponds to a NNW–SSE-trending depocentre that originated during the late Carboniferous–Permian (Lesta et al. 1980). Such a depocentre has apparently transtensive features and may have extended into the Permo–Triassic, allowing the granite intrusions of the Somuncurá and Deseado massifs (Sylwan et al. 2011), shown in Fig. 1.3.

During the Middle Jurassic in Patagonia, there occurred an extensional geotectonic process, in which the depocentres with a half-graben geometry were filled with volcaniclastic, lacustrine, and marine sediments in a late-rift tectonosedimentary stage (Sylwan et al. 2011). These half-grabens configure the depositional scenario of the Neocomian Group in the stage mentioned.

In the course of this period, a S-trending extensional fault zone, later affected during the Tertiary by compressive tectonic activity with effects of tectonic inversion, is responsible for the present-day fold belt (Bernardides, southern end of the Patagonides).

Towards the east, W–E-trending extensional faults occur, extending until the opening of the Atlantic Ocean (Fitzgerald et al. 1990). It was followed by a tectonic phase, which was responsible for an erosive event that affected the entire basin. This created available space for a new sedimentary cycle, represented by the Chubut Group (Lesta et al. 1980), with a W–E trend and corresponding to the thermal subsidence phase (sag) of the basin (Sylwan et al. 2011).

The extensional and transtensional episodes were different from those of the previous cycle, and they continued from the latest Lower Cretaceous until the Palaeogene, in retroarc conditions. Then the subsidence rate of the basin accelerated, coinciding with the ingression of the Salamanca Formation (Palaeocene) and the onset of a new Tertiary sedimentary cycle (Figari et al. 2002). These extensional conditions changed, with the uplifting of the Palaeocene fold belt being the result of the transpressure and tectonic inversion related to the development of the Andes.

Since the Eocene, the basin has behaved as a broad tectonic basin with moderate subsidence and four marine ingressions, alternating in the filling with fine-grained pyroclastic rocks that originated in eruptive centres located to the west (Legarreta et al. 1990).

Even though the basin shows, as a result of the dominant extensional tectonics, major E–W-trending structural features, the fold belt (Bernardides)—which is elongated in a N–S trend—makes it possible to divide it into two sectors, east and west. This may also be applied in the former, where areas referred to as "Northern Flank", "Southern Flank", "Basin Centre", and "Offshore" (Fig. 3.1) can be recognized.

3.3 Geomorphology, Surface Hydrology and Soils

Out of the major positive landforms of the Extra-Andean Patagonia mentioned in the first chapter, only the stepped tablelands (Northern Patagonian Tablelands) and the Patagonides occur within the basin, since the two extensive massifs (Somuncurá and Deseado massifs) are located out of the basin, framing it.

Among the notable positive units are (a) the Northern Patagonian Tablelands, (b) the Bernardides (southern end of the Patagonides), (c) low hills, (d) littoral ridges, and (e) inland dunes.

The negative units include (a) the depression of the central lakes, (b) the valley of the central lakes and its tributaries, (c) the Chico River valley, (d) the Deseado River valley, (e) the Chubut River valley, and (f) endorheic depressions.

The Northern Patagonian Tablelands are widely distributed over the central and eastern sectors (Fig. 3.9). In the former, basaltic tablelands with lava flows and volcanic necks predominate, with the flows corresponding to alkaline plateau basalts (Ramos 1999).

Towards the east, raised plains stand out, decreasing in a stepped manner towards the sea, overlain by Patagonia Formation or fluvio-glacial gravels with a wide areal extent, known as "Rodados Patagónicos" (Patagonian Shingle

Fig. 3.9 Punta Maqueda tableland, which ends in the San Jorge Gulf. View to the north of Caleta Olivia. Courtesy of Rufino Sánchez

Formation, Fidalgo and Riggi 1970) (Fig. 3.10). These tablelands are also locally known as "*pampas*" and "*mesetas*", such as the Pampa de Salamanca, Pampa del Castillo, Meseta Cuadrada, Pampa de los Guanacos, Meseta de Montemayor, and Meseta Pelada.

The Bernardides constitute a series of hill units in the southern part of the Patagonides, to the north of the Senguerr River, characterized by anticlines separated by N–NW-trending faults and formed by the tectonic inversion of half-grabens sensu Homovc and Lucero (2002) (Fig. 3.11).

The oldest lithology exposed in the Sierra de San Bernardo dates from the Lower Cretaceous and the environment includes Sierra de San Bernardo basalt flows (Ramos 1999). The hill successions are the Sierra de San Bernardo, Sierras de Corrientes, Sierra de Cañadón Grande, and Sierra Nevada, among others, and the maximum heights are the Sierra Nevada (1558 m a.s.l.) and the Cerro Colorado (1371 m a.s.l.).

Low hills, with various trends related to different post-Jurassic episodes, occur in different locations, generally emerging from the tablelands (Sierra Negra, Sierra Chaira, Sierra Cuadrada, Cerro Tres Picos and Ovejas Grandes) with low heights. They are also erosion remnant (inselberg) as show in Fig. 3.12.

The littoral ridges, generally composed of gravel and sand, occur along the sea coast of the gulf. Inland dunes can also be found in different locations within the

Fig. 3.10 Outcrop of the Patagonia Formation on a tableland on the coast of the San Jorge Gulf. Courtesy of Rufino Sánchez

Fig. 3.11 View of the Bernardides, west of Sarmiento valley. Courtesy of Rufino Sánchez

Fig. 3.12 Erosion remnant (inselberg). Cerro Pan de Azúcar to the south of Comodoro Rivadavia City. Courtesy of Rufino Sánchez

territory, in the western section of the Bernardides, the northwest of the Province of Santa Cruz and the eastern edge of some endorheic depressions (Fig. 3.13).

Among the negative landforms, the one that stands out due to its size and occurrence is the depression of the central lakes, taken up by the lakes Musters and Colhué Huapi, which are the largest extra-Andean lake bodies in the Province of Chubut. They occur in a depression of aeolian origin with alluvial-aeolian deposits (Gran Bajo de Sarmiento) that then may have constituted a single lake before its present-day conformation (González Díaz and Di Tomasso 2014).

The Senguerr River valley, with an original W–E trend, similar to the trend of its Andean tributaries, then shifting to the N–S and finally to the SW–NE, is a typical watercourse with montane headwaters, a well-defined valley and good slope until its middle course. Its morphology and trajectory then become meandering up to the lower course, where the slope decreases significantly, while the meandering increases, as does the width, reaching 60–70 m and then flowing into the Musters Lake.

The morphology of the Chico River is different: it was the former drainage of the above-mentioned lakes and at present has a transitory behaviour (dry most of the year), as will be explained below. With a linear course, small ephemeral tributaries, and a broad alluvial valley that reaches a development of up to 2.5 km, the Chico River has a SW–NE trend, discharging into the Chubut River 15 km upstream from

Fig. 3.13 View of an active dune in Las Flores zone, 10 km to the southeast of Colhué Huapi Lake. Courtesy of Julio Cotti Alegre

the Florentino Ameghino Dam. Its mouth on the Colhué Huapi Lake is practically obstructed by a growing silty–sandy bar.

The Deseado River has Andean headwaters, at Ap Iwan Hill (2037 m a.s.l.), as Fénix Grande River. Its upper valley is typical of a perimontane river, with a trapezoidal profile and steep slopes up to the Pinturas River, a tributary on the right margin. From that point, it widens considerably, until its alluvial plain reaches a width of 5000 m, and its pattern becomes meandering, with a transitory behaviour. Its tributaries—Fénix Grande River, Fénix Chico River, Arroyo Page, Cañadón Grande, Cañadón Botello and Cañadón Primavera—have small basins and a dendritic pattern; the Pinturas River is the main affluent. It flows into the Atlantic Ocean as a ria, out of the basin, after running over 615 km.

As regards the Chubut River, only its middle section is within the San Jorge Gulf Basin (Valle de Los Altares). It still has the typical morphology of mountain rivers, despite the fact that at that point it crosses the tablelands, almost with no affluents, between the Somuncurá Massif and the San Jorge Gulf Basin, forming a deep canyon carved into volcanic rocks.

The endorheic depressions vary in size and shape, though they are generally subcircular to ovoid. The most important one is the Gran Bajo Oriental, to the southwest of Comodoro Rivadavia, with an aeolian origin by deflation, very flat, oval-shaped with an irregular outline, a height of 76 m a.s.l., and an entirely centripetal character.

Concerning the surface hydrology, three main courses occur within the context of the basin: the Senguerr and Chico rivers, which mostly flow within it, and the Chubut River, which only runs through a reduced marginal sector.

The Senguerr River, with a length of 350 km, is a perennial watercourse with a pluvio-nival regime and it loses water to groundwater. It springs from the Fontana Lake at a height of less than 1000 m a.s.l. and its rushing waters are initially hemmed in by a geoenvironment resulting from glacial abrasion, subsequently flowing through a broad fluvial valley (Moyano 2016).

When the river reaches the tablelands, it divides into distributaries and it gradually acquires a braided to meandering pattern, reaching the southern end of the Bernardides with a N–S trend, which suddenly shifts to a SW–NE at the bend known as Codo del Senguerr. It reaches the Musters Lake, where at present it discharges into a broad alluvial valley, whereas 70-years ago it connected to the Colhué Huapi Lake by means of an arm called False Senguerr, and from there continued under the name Chico River, reaching the course of the Chubut River (Valladares 2004).

This continuity ended in 1939, due to a decrease in rainfall, poor management in the upper basin and as a result of the occurrence of a bar at the point in which it discharges into the Chico River, with the exception of some isolated moments of ephemeral behaviour. The consequence is the progressive desiccation of the Colhué Huapi Lake (Fig. 3.14).

The river discharge is 48.6 m^3/s and its regime is pluvio-nival, with a peak flow in winter due to rainfall and another one in spring (October–November), resulting from the thawing of Andean snow (Subsecretaría de Recursos Hídricos 2002). The extreme average daily flow rates were 283.4 m^3/s (1997–1998) and 2.5 m^3/s (1967–1968); all of these values were registered at the flow rates weather station, located near the Codo del Senguerr.

Fig. 3.14 Senguerr River from Codo del Senguerr. In the background to the right, end of the Bernardides, Courtesy of Julio Cotti Alegre

During the coldest winter months, the river usually freezes and remains frozen for more than 1 or 2 months, acting like a weir until the first liquid contributions break the ice bridge, which may lead to the waterlogging of the Senguerr River in the Sarmiento Valley, which is in the lake area. It is important to take into consideration that, even though in the tablelands the rainfall does not exceed 200–300 mm/yr, at headwaters the annual mean is of the order of 1600 mm/yr and the influx of discharge during snowmelt is very significant.

On the left bank of the upper basin, the Senguerr River receives contributions from the tributary streams Arroyo Genoa and Arroyo Shaman-Apeleg, whereas on the right bank from the Arroyo Verde. In the middle basin, the main inflow is from the Mayo River, which is perennial with glacial sources, as well as from its tributaries, the Arroyo Chalía and Río Guenguel, and numerous minor watercourses until after the Codo del Senguerr.

The Chico River, an intermittent stream, is the natural outlet—though cut off at present—of the surplus of the Colhué Huapi Lake (Valladares 2004). The width of its streambed accounts for the historical relevance this river has had as a tributary of the Chubut River. When discussing above the Senguerr River and the mechanism of the central lakes, the causes of its drying were discussed. Its length reaches 300 km and several ephemeral gullies join it on both margins.

The Deseado River has a transitory behaviour, running intermittently downstream from the point where its main tributary, the Pinturas River, flows in; up to that point, the annual discharge is 5 m^3/s. The regime is pluvio-nival, with a peak during the autumn and winter rainfall, and another during the spring thaw. Its tributaries were mentioned above, where its valley is described.

Finally, as regards the Chubut River, approximately 130 km out of its 810 km go through a limited, marginal section of the territory in the northern boundary, after crossing the entire Province of Chubut, which it enters in the northwest corner and then discharges into the Atlantic Ocean. Within the San Jorge Gulf Basin, it runs through a canyon (Valle de los Mártires, Valle de Los Altares, and Valle Paso de Indios). Its sources are in the Andes and its regime is pluvio-nival, with peak flows in July–August (pluvial) and October–November (nival), extremely close in time, and with minimum flow from November to April. Its discharge reaches 57.3 m^3/s, with maximum and minimum instantaneous discharge rates of 372 and 4 m^3/s, respectively. Due to its position, its exploitation in the basin under study is extremely difficult.

In the basin, *soils* belong to the Aridisol order, which—as the name implies—are from arid regions, pale in colour, poor in organic matter, with low pedogenic development and a low decomposition rate. An illuvial calcic horizon can be recognized, and generally another argillic horizon, supporting Xerophytic vegetation, sometimes halophytic, fit for the grazing of livestock.

Among the great groups within this order, Natrargids occur in the flattest and/or most depressed sectors in the central-northern area and in the plains that skirt the Deseado River valley, to the south. Haplargids are found in the gently undulating tablelands to the southwest and centre-south, whereas in the central undulating plains, to the west of the Chico River and in the coastal terraces of the gulf, practically around its entire perimeter, Paleargids occur. Over the flood plain of the Senguerr River and in the sources of the Chico River, Petrocalcids can be recognized.

Soils of the Entisol order are typical of the slopes of the valleys of the Senguerr-Chico rivers and the Deseado River (Torriorthents), of depressions in alluvial plains (Torrifluvents), and of beaches of the Buenos Aires Lake (Fluvaquents).

Mollisols can be found in the Chico River valley: Calcixerolls in intermontane tablelands and plains, and Haploxerolls in the alluvial plain; the latter also occur in low plains. Argixerolls occupy a coastal area that extends throughout the entire coastal region to the north of the gulf.

On the western boundary, skirting the Andes, occurrences of Entisols (Haplocryolls, Cryaquolls, and Haploxerolls) can be observed in landforms connected to the Glacial landscape, valleys and fluvio-glacial plains. Another great group of Entisols, of limited occurrence, is that of the alluvial complexes of Endoaquolls, which occur in the narrow alluvial plain of the Chubut River (Cruzate 2011; Cruzate et al. 2011).

As regard the land use capability classification (USDA 1961), the vast majority of them correspond to classes V and VI, that is to say, with a limited grazing capacity, in this particular case, mostly for sheep farming. The limitations are obviously due to the climatic characteristics, with a scarce precipitation regime, materials with low to null water retention capacity, and organic matter deficiency. It is only in the valleys of the Senguerr River and its tributaries, of the Chubut River, and sectors of the Deseado River upper basin that class II–IV soils occur. In turn, in the peri-Andean area, there are sectors with a class VII capacity, exclusively suited for forest use.

The regions with soils fit for grazing are recovering from sheep overpopulation that they were subject to until the 1970s. On the western boundary and within the upper basin of the Senguerr River, cattle farming has increased at the expense of the creation of artificial wetlands, due to the waterlogging of sectors of the valleys of the rivers Genoa, Shaman, Apeleg, and Chalía, among others. Attempts are made to imitate the natural wetlands, locally known as *mallines*, with grass vegetation and permanent water accumulation.

This practice has introduced a serious distortion in the surface and groundwater regime, causing among other problems the above-mentioned loss of continuity in the Senguerr-Chico fluvial system, to the point that it has practically become endorheic.

The most important hydrological role of soils in this region is given by their texture, grain size, and specific retention capacity. The predominance of coarse textures in the parent materials and the extremely low specific retention, which are

Fig. 3.15 Xerophytic vegetation with low density cover on Aridisols soils. Bajo Barreal Formation. Courtesy of Julio Cotti Alegre

general characteristics of Aridisols and Entisols, facilitate infiltration; both of these characteristics will be discussed once again when describing the recharge mechanisms. Only part of the Mollisols do not possess this quality (Fig. 3.15).

3.4 Geology

As anticipated in Sect. 3.2, when summarizing the geological history of the basin, the depositional scenario consisted in depocentres with half-graben geometry, filled during a late-rift stage with sediments of the Neocomian Group, whose most permeable layers host the aquifers that compose the geohydrological system and the oil and gas reservoir rocks.

Given that the main objective of this book is to disseminate the hydrogeological research undertaken in the region and that its secondary objective is to discuss its relationship with HC production, the geological perspective focuses precisely on the rocks containing both fluids and their lithological characteristics, which is why only a brief mention of the pre-Jurassic and Jurassic components will be made.

The discretization into sectors classically referred to as "Eastern Sector"—subdivided into "Northern Flank", "Basin Centre", and "Southern Flank"—and "Western Sector", as well as the "Bernardides" and "Offshore" sectors, was adopted following Homovc and Lucero (2002) (Fig. 3.1).

According to Sylwan et al. (2011), the Somuncurá and Deseado massifs—which define this intracratonic basin:

> are composed of a heterogeneous group of rocks that includes metamorphic rocks and intrusives (lower Palaeozoic–Precambrian), granites and schists (Devonian), sedimentary units (Carboniferous–Permian), igneous rocks (Permian–Triassic), pelites, psammites and pyroclastites (Triassic), sedimentary, volcaniclastic and marine rocks and their continental equivalents (Lias) and the rocks of the Volcano-Sedimentary Complex (Middle to Upper Jurassic).

The Precambrian–early Mesozoic igneous rock units, as well as other late Palaeozoic and Mesozoic sedimentary rocks, underlie the Upper Jurassic sedimentary column, which has a maximum thickness of 8000 m in the centre of the basin and is in general dominated by Cretaceous and Tertiary continental sediments (Sylwan et al. 2011). In many cases, the Cretaceous sediments are in direct contact with the underlying pre-Jurassic sediments mentioned above.

The Upper Jurassic Volcano-Sedimentary Complex (Clavijo 1986), which is composed of volcanic and pyroclastic rocks, consists of lithostratigraphic units that occur in different sectors, such as the Lonco Trapial Group (Northern Flank, Lesta and Ferello 1972) and the Bahía Laura Group (Southern Flank, Feruglio 1949). The former is represented by basalts and other volcanic and volcaniclastic rocks, whereas the latter gathers the Chon Aike Formation (ignimbrites, rhyolites, porphyrites and tuffs) and the La Matilde Formation (tuffs, sandstones, and conglomerates), generally overlying the former group. The complex is probably of a Middle and Upper Jurassic age.

The Cretaceous period starts with infill deposits of the Neocomian Group in the half-grabens developed during the Jurassic, outcropping in the Western Flank, where they reach their maximum thickness, even though they are not as well known as in the western area, where they were affected by the drilling for oil exploration. The Pozo Anticlinal Aguada Bandera Formation, of lacustrine origin, and the Pozo Cerro Guadal Formation, of estuarian-lacustrine origin, have been recognized here. The Pozo Anticlinal Aguada Bandera Formation, with a development that ranges from 600 to 5000 m, is composed of black and dark grey claystones, siltstones and mudstones intercalated with fine-grained sandstones, generally overlain by coarse-grained clastic sediments in the thicker sequences (Fitzgerald et al. 1990). Even though their origin is lacustrine, marine deposits have been assigned to this unit in the west of the basin.

The Pozo Cerro Guadal Formation overlies the above-mentioned unit with angular discordance: it is composed of quartz sandstones, hard and compact, with a tuffaceous matrix, tuffaceous siltstones, clear tuffs, and black pelites (Lesta and Ferello 1972). Its thickness in the type location, where it has been completely drilled through, is 560 m, with an average of 300 m. Although its origin is lacustrine like the above-mentioned formation, it has a significantly lower organic content (Sylwan et al. 2011).

These deposits of the Neocomian Group thin eastward, displaying a marginal character (Figari et al. 2002), and underlie a cycle of major HC significance that includes all of the rest of the Cretaceous deposits, that is, the Chubutian Cycle or Chubut Group.

This group includes four units that are given different names depending on the sector of the basin, whether it is the Eastern Sector in its Northern or Southern flanks, or the Western Sector, which is why a composite stratigraphic scheme where the equivalences are established is given (Table 3.1).

Taking as reference the Eastern Sector, they are: the Pozo D-129 Formation is equivalent to the Matasiete Formation in the Western Sector, the Mina del Carmen Formation to the Castillo Formation, the Comodoro Rivadavia-Cañadón Seco Formation is equivalent to the lower member of the Bajo Barreal Formation, and the Yacimiento El Trébol-Meseta Espinosa Formation to the upper member of the Bajo Barreal Formation in the Western Sector (Lesta et al. 1980; Uliana and Legarreta 1999; Homovc and Lucero 2002), Fig. 3.16.

The Pozo D-129 Formation is present in practically the entire basin, and it is composed of deposits of deep lacustrine, shallow lacustrine, fluvial and deltaic origin (Hechem et al. 1987). Pelites predominate especially in the first case, whereas in the others, silts, sandy silts, fine-grained sandstones, and even conglomerate sandstones can be found. Its thickness is of 800 m in the type location borehole (Diadema Argentina), varying between 300 and 1500 m. The pelites have a high organic content, which emphasizes their character as a HC source rock, probably the main one in the basin. To the west of the Bernardides, its equivalent Matasiete Formation occurs (Lesta and Ferello 1972) and it is represented by fluvial and pyroclastic sediments.

The sequence continues with the Mina del Carmen Formation, characterized by pyroclastic deposits (tuffs) and pelites, with the presence of fluvial and fluvio-deltaic

Table 3.1 Stratigraphic scheme. Composed from Malumian (1999), Homovc and Lucero (2002), and Uliana and Legarreta (1999)

PERIOD	GROUP	FORMATION	THICKNESS
TERTIARY		*Santa Cruz Fm.*	80–200 m
		Patagonia Fm.	50–500 m
		Sarmiento Fm.	100–250 m
		Río Chico Fm.	100–300 m
		Salamanca Fm.	100–150 m
UPPER CRETACEOUS	CHUBUT	*Bajo Barreal Fm.* (upper member) = *Yac. El Trébol Fm.* = *Yac. Mes. Espinosa*	700–1000 m
		Bajo Barreal Fm. (lower member) = *C. Rivadavia Fm.* = *Cañadón Seco Fm.*	700–1000 m
LOWER CRETACEOUS		*Castillo Fm.* = *Mina del Carmen Fm.*	400–1500 m
		Matasiete Fm. + Pozo D-129 Fm.	900–2500 m
	NEOCOMIAN	*Pozo Cerro Guadal Fm.* *Pozo Anticlinal Aguada Bandera Fm.*	100–1500 m
JURASSIC	BAHÍA LAURA	*La Matilde Fm.* *Chon Aike Fm.*	> 500 m

Fig. 3.16 Castillo Formation in El Molino (Senguerr River area). Courtesy of Hidroar SA

lithologies towards the flanks (sandstones and siltstones). Towards the west, it crops out as the Castillo Formation (Lesta and Ferello 1972), with tuffaceous sandstones, compact tuffs and thin sandstones of discontinuous occurrence. The psammitic facies range from predominating to being almost absent, showing an irregular distribution. This unit reaches thicknesses of up to 2000 m along the axis of the basin and of only 220–300 m at its edges, with different authors assigning them an old age, within the Early Cretaceous (Aptian–Albian) sensu Fitzgerald et al. (1990) and Albian sensu Figari et al. (2002) in Sylwan et al. (2011). Whereas the Pozo D-129 Formation is a typical oil and gas source rock, the Mina del Carmen and Castillo formations are important reservoir rocks.

These units are overlain by the equivalent formations Comodoro Rivadavia (Eastern Sector, Northern Flank), Cañadón Seco (Eastern Sector, Southern Flank), and the lower member of the Bajo Barreal (Western Sector), Fig. 3.17. Their lithological composition is characterized by the presence of white lithic tuffs stratified in thin beds, greyish–white sandstones, well rounded conglomerates with quartz and volcanic clasts, tuffaceous sandstones, and reddish and yellowish pelites that usually contain coarse-grained sandstone lenses.

In the Northern Flank, the sandstone beds are thicker and they develop more frequently than in the Southern Flank, with greater amounts of sand than in the Chubut Group units described above, which are controlled by faults (Hechem 1998). Their thickness varies between 300 m in the flanks to 1200 m in the centre of the basin. The age, despite discrepancies, could be estimated in the lower levels

Fig. 3.17 Bajo Barreal Formation covered by the Laguna Palacios Formation in Sierra de San Bernardo. Courtesy of Rufino Sánchez

of the Late Cretaceous. Its practical importance lies in the fact that it contains important conventional HC reservoirs.

The Chubutian segment in the stratigraphic column ends with the units present in the Eastern Sector of the reservoir, that is, the El Trébol and Meseta Espinosa formations (in the Northern and Southern flanks, respectively), and the upper member of the Bajo Barreal Formation in the Western Sector, where it crops out. They deposited during an expansion episode with subsidence reactivation, in finer facies than the underlying deposits (Sylwan et al. 2011). They are mainly pelitic, with the presence of sands in the Northern Flank. In the Western Sector, the Laguna Palacios Formation appears laterally, composed of pyroclastic material with the frequent development of conglomerate sandstones and palaeosols (Lesta and Ferello 1972). The thickness of the units ranges from 200 to 250 m at the edges to 800 m along the axis of the basin and it is assigned a Maastrichtian age, corresponding to the cuspidal portion of the Late Cretaceous (Fig. 3.18).

A brief reference to the Offshore sector, taken from Sylwan et al. (2011), indicates the presence of main and secondary faults, which originate a somewhat different geometry in the Cretaceous deposits (Neocomian and Chubutian cycles), within a context that is generally similar to the one in the Eastern Sector, over a basement of Palaeozoic metamorphic and plutonic rocks, as well as Jurassic pyroclastic and volcanic rocks.

Fig. 3.18 Opalized fossil tree remains. Sarmiento Petrified Forest (formerly, José Ormachea Petrified Forest) in early Palaeogene sections. Courtesy of Julio Cotti Alegre

Discordantly overlying the Cretaceous lithologies—paraconcordantly in certain sectors—the Tertiary formations of the Palaeocene (Danian) and probably of the final stages of the Maastrichtian (Malumian 1999) can be found. The cycle begins with the Salamanca Formation (= San Jorge Formation), which is the first Atlantic marine ingression into the basin. Its contents are sandstones, fragmented pelites, and conglomerates, increasing in grain size towards the west, with a thickness at its base composed of glauconitic sandstones that constituted the first oil field discovered in Argentina in 1907. Its thickness gradually decreases towards the west.

The Río Chico Formation begins a continental cycle, which extends to almost the whole extent of the basin. It consists of upper Palaeocene bentonite clays, fine tuffs, tuffaceous sandstones and conglomerates, with the presence of phosphate contents. It reaches thicknesses of between 150 and 250 m, and there has been a tendency to define it as Río Chico Group, composed in turn by four formations (Uliana and Legarreta 1999). It is overlain in erosive discordance by the ash-grey to white ash tuffs of the Sarmiento Formation, of Eocene–Oligocene age (Mazzoni 1985). Due to their wide geographic distribution and frequent outcrops in the Eastern Sector (Fig. 3.19), they are easily distinguishable by their colouring and appearance, with organ pipes and sometimes also with the presence of several basalt flows.

Slightly discordantly, or at times concordantly, over the Río Chico Formation occur the deposits of the Sarmiento Formation, whose lithology is composed of

Fig. 3.19 Badlands (Río Chico Formation) in the Anticlinal Grande area, 22 km to the southeast of Colhué Huapi Lake, in Oilflied area. Photo by Mario Hernández

fine-grained cineritic and bentonitic tuffs, and ash flows. The latter formation consists of thick friable beds of pelitic pyroclastic rocks that are whitish, greyish, or slightly yellowish in colour, and secondarily associated with greenish grey of mixed composition of bentonite and volcanic ash with intercalations of gypsum and sandstones of volcanic origin. Based on the occurrence of fossil mammals these rocks have been assigned to the lower Oligocene. Originally, they were referred to as "Piroteriense" or "Piroteriano", whereas some authors prefer the term "Grupo Sarmiento", which in turn is composed of four formations (Ardolino et al. 1999). In the upper portion, basalt flows occur (Fig. 3.20).

The Patagonia Formation (Fig. 3.20) is composed of deposits of the late Oligocene–Miocene marine transgression, the most important one after the Salamancan Formation. Originally, in the 19th century, it was divided into the Julian (lower), Leonian, and Superpatagonian levels. Its lithology generally comprises sandstones intercalated with claystones and siltstones, including tough coquina banks. Historically, it has experienced successive and sometimes cyclical changes in allocation and name (Hernández and Hernández 2013): the San Julián Formation corresponds to the Juliense level, or the Monte León Formation to the Leonense level. It has also been proposed to regard it as a group, with the above-mentioned formations as components, or in other cases to name it Chenque Formation (Bellosi 1990; Paredes 2002).

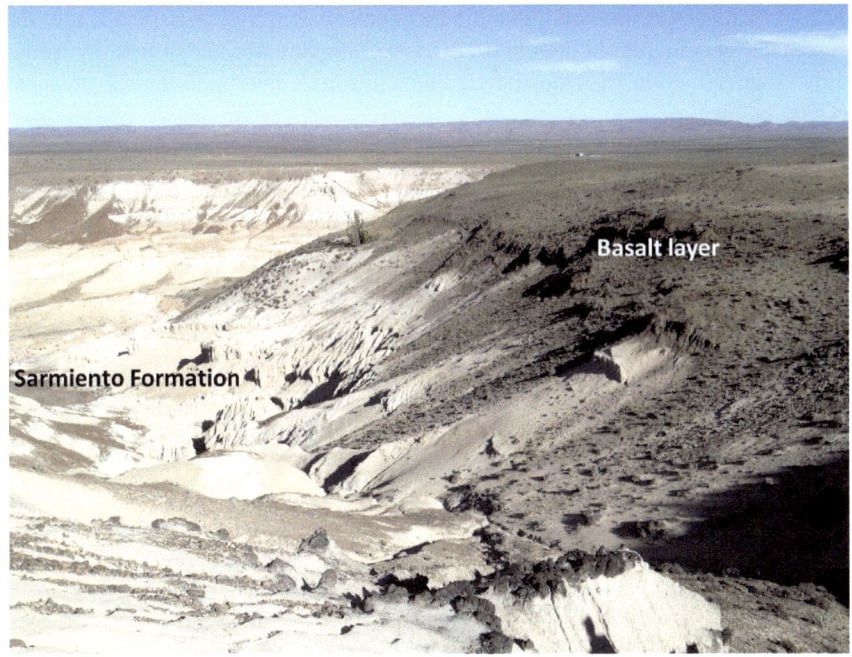

Fig. 3.20 Basalts overlying the Sarmiento Formation at Anticlinal Grande. Courtesy of Julio Cotti Alegre

Its genesis corresponds to the greatest depth of the Patagonian Cenozoic seas (Malumian 2002), with great biogenic silica contribution. Thus, the pelites that originated in the tuffs of the Sarmiento Formation were eroded and redeposited in a marine environment. Depending on the palaeomorphological position, such pelites contribute to important clayey–silty-clayey thicknesses at the base of the formation. On the other hand, in other sectors, sandy sedimentary rocks predominate and the finer fractions are subordinate, normally as intercalations.

The Santa Cruz Formation is part of a sedimentary complex together with the upper section of the Patagonia Formation, with a succession of fine- to medium-grained sandstones, tuffaceous claystones, claystones, tuffs and tuffites, generally assigned to the Miocene (Nullo and Combina 2002). Normally, it is very difficult to differentiate both formations in well logs, except on the basis of their microfossil content (Hernández and Hernández 2013).

Finally, the most widely disseminated post-Tertiary materials are the ones known informally as "Rodados Patagónicos", generally coarse vulcanitic and porphyric gravels of Andean origin. They are polygenetic, with a fluvio-glacial genesis being mostly recognized. They cover the tablelands and tend to originate terraced deposits, such as the ones of the Deseado River, the Senguerr River and other minor watercourses. They have had several formal names, such as the La Avenida Formation, Cordón Alto Formation, Pampa de la Compañía Formation and Mata Grande Formation. Their age is probably Pleistocene.

The Tertiary units described are of special interest for the purposes of this book, since their characteristics will allow the definition of the physical component of the conceptual model representing the behaviour of the groundwater system, by means of the transformation of lithostratigraphic units into hydrolithological units, as detailed below.

References

Ardolino AA, Franchi M, Remesal M, Salani F (1999) El volcanismo en la Patagonia Extrandina. In: SEGEMAR (ed) Geología Argentina Anales: Instituto de Geología y Recursos Minerales, Buenos Aires, vol 29, no 2, pp 579–612

Bellosi ES (1990) Formación Chenque: registro de la transgresión patagoniana en la Cuenca San Jorge. XI Congreso Geológico Argentino, Actas 2:57–60 (San Juan)

Burgos JJ, Vidal AL (1951) Los climas de la República Argentina según la nueva clasificación de Thornthwaite. Meteoros 1(1):3–32, Servicio Meteorológico Nacional, Buenos Aires

Clavijo R (1986) Estratigrafía del Cretácico Inferior en el sector occidental de la Cuenca del Golfo San Jorge. Boletín de Informaciones Petroleras 9:15–32 (Buenos Aires)

Cruzate GA (2011) Suelos y Ambiente de Santa Cruz. Instituto de Suelos, Instituto Nacional de Tecnología Agropecuaria (INTA), Buenos Aires

Cruzate GA, Panigatti JL, Moscatelli G (2011) Suelos y Ambiente de Chubut. Instituto de Suelos, Instituto Nacional de Tecnología Agropecuaria (INTA), Buenos Aires

Feruglio E (1949) Descripción geológica de la Patagonia. Dirección General de Yacimientos Petrolíferos Fiscales 1:2 (Buenos Aires)

Fidalgo F, Riggi JC (1970) Consideraciones geomórficas y sedimentológicas sobre los Rodados Patagónicos. Revista Asociación Geológica Argentina 25(4):430–443 (Buenos Aires)

Figari EG, Strelkov E, Cid de la Paz MS, Celaya J, Laffitte G, Villar H (2002) Cuenca del Golfo San Jorge: Síntesis estructural, estratigráfica y geoquímica. In: Haller MJ (ed) Geología y Recursos Naturales de Santa Cruz. Relatorio del XV Congreso Geológico Argentino. El Calafate, vol III-1, pp 571–601

Fitzgerald MG, Mitchum RM Jr, Uliana MA, Biddle KT (1990) Evolution of the San Jorge Basin, Argentina. Bull Am Assoc Petroleum Geologists 74(6):879–920

González Diaz EF, Di Tommaso I (2014) Paleogeoformas lacustres en los lagos Musters y Colhué Huapi, su relación genética con un paleolago Sarmiento previo, centro-sur del Chubut. Revista Asociación Geológica Argentina 71(3): 416–426 (Buenos Aires)

Hechem JJ (1998) Arquitectura y paleodrenaje del regional efímero de la Formación Bajo Barreal, Cuenca del Golfo San Jorge, Argentina. Boletín de Informaciones Petroleras 53:21–27 (Buenos Aires)

Hechem JJ, Figari EG, Musacchio EA (1987) Cuenca del Golfo San Jorge. Hallazgo de la Formación D-129. Información estratigráfica y paleontológica. Petrotecnia 28(11):13–15 (Buenos Aires)

Hernández MA (2000) Estudio geohidrológico de la región Cerro Rubio-Cerro Vanguardia, provincia de Santa Cruz. Unpublished doctoral thesis, Facultad de Ciencias Naturales y Museo, Universidad Nacional de La Plata, 163 pp

Hernández L, Hernández MA (2013) Características hidrolitológicas de las formaciones Patagonia y Santa Cruz. Cuenca del Golfo San Jorge. (Provincias de Chubut y Santa Cruz). In: González N, Kruse EE, Trovatto MM, Laurencena P (eds) Agua subterránea recurso estratégico. EDULP, La Plata, vol I, pp 112–117

Homovc JF, Lucero M (2002) Cuenca del Golfo San Jorge: Marco geológico y reseña histórica de la actividad petrolera. In: Rocas reservorio de las cuencas productivas de la Argentina. Instituto Argentino de Petróleo y Gas (IAPG), Buenos Aires, pp 119–126

INDEC (2011) Censo Nacional de Población Hogares y Viviendas 2010. Instituto Nacional de Estadística y Censos (INDEC), Buenos Aires

Legarreta L, Uliana MA, Torres MA (1990) Secuencias deposicionales cenozoicas de Patagonia central: sus relaciones con las asociaciones de mamíferos terrestres y episodios marinos epicontinentales. Evaluación preliminar. II Simposio del terciario de Chile, Actas. Concepción, Chile, pp 135–176

Lesta PJ, Ferello R (1972) Región Extraandina de Chubut y Norte de Santa Cruz. In: Leanza AF (ed) Primer Simposio de Geología Regional Argentina, Academia Nacional de Ciencias de Córdoba, pp 601–653

Lesta PJ, Ferello R, Chebli G (1980) Chubut extrandino. In: Turner JCM (coordinator): Segundo Simposio de Geología Regional Argentina. Academia Nacional de Ciencias de Córdoba, vol 2, pp 1307–1387

Malumian N (1999) La sedimentación en la Patagonia Extraandina. In: SEGEMAR (ed) Geología Argentina Anales. Instituto de Geología y Recursos Minerales, Buenos Aires, vol 29, no 18, pp 557–612

Malumian N (2002) El Terciario marino. Sus relaciones con el eustatismo. In: Haller MJ (ed) Geología y Recursos Naturales de Santa Cruz. Relatorio del XV Congreso Geológico Argentino. El Calafate, I-15:237–244 (Buenos Aires)

Mazzoni MM (1985) La Formación Sarmiento y el vulcanismo paleógeno. Revista de la Asociación Geológica Argentina 40:60–68 (Buenos Aires)

Moyano CH (2016) Régimen hídrico y morfología fluvial de la cuenca del río Senguerr. Cátedra Geomorfología Facultad de Humanidades—Universidad Nacional de Catamarca, San Fernando del Valle de Catamarca, 11 pp

Nullo FE, Combina AM (2002) Sedimentitas terciarias continentales. In: Haller MJ (ed) Geología y Recursos Naturales de Santa Cruz. Relatorio del XV Congreso Geológico Argentino. El Calafate, I–16:245–258 (Buenos Aires)

Paredes JM (2002) Asociaciones de facies y correlación de las sedimentitas de la formación Chenque (Oligoceno-Mioceno) en los alrededores de Comodoro Rivadavia, cuenca del Golfo San Jorge, Argentina. AAS Revista. Asociación Argentina de Sedimentología. Buenos Aires, vol 9, no 1, pp 53–64

Ramos V (1999) Las provincias geológicas del territorio argentino. In: SEGEMAR (ed) Geología Argentina, Anales. Instituto de Geología y Recursos Minerales, Buenos Aires, vol 29, no 3, pp 41–96

Subsecretaría de Recursos Hídricos (2002) Atlas Digital de los Recursos Hídricos Superficiales de la República Argentina, Buenos Aires [CD-ROM]

Sylwan C, Droeven C, Iñigo J, Mussel F, Padva D (2011) Cuenca del Golfo San Jorge. VIII Congreso de Exploración y Desarrollo de Hidrocarburos. Simposio Cuencas Argentinas: visión actual. Instituto Argentino de Petróleo y Gas (IAPG), Mar del Plata, pp 139–183

Thornthwaite CW, Mather JR (1955) The water balance. Drexel Institute of Technology, Laboratory of Technology, Publications in climatology vol 8, No 1 Centerton, NJ 104 pp

Uliana MA, Legarreta L (1999) Jurásico y Cretácico de la cuenca del Golfo San Jorge. In: SEGEMAR (ed) Geología Argentina, Anales. Instituto de Geología y Recursos Minerales, Buenos Aires, vol 29, pp 496–510

USDA (1961) Land capability classification. U.S. Departament of Agriculture Handbook 210

Valladares A (2004) Cuenca de los ríos Senguerr y Chico. Cuenca No. 66. Instituto Nacional del Agua, Buenos Aires

Chapter 4
Geohydrology

Abstract The state-of-the-art hydrogeological research, the chronology of studies on the subject, and the conformation of the physical component of the geohydrological system are analyzed. Two units are observed: the upper (SGS) and lower geohydrological system (SGI), from the Salamanca Formation to the surface. In the SGS the main aquifers occur (Patagonia and Santa Cruz formations), generically called Patagonian aquifer. They may be phreatic or semi-confined, with marked anisotropy and heterogeneity, an 8–15 m³/h yield, specific yields of 100–150 m³/d m, and a transmissivity of over 80 m²/d. The vadose zone reaches thicknesses of over 20 m in tableland areas. The geohydrological cycle is described. Recharge may be direct autochthonous, indirect autochthonous, or allochthonous. Circulation was interpreted by making an equipotential diagram: the general flow direction is towards the east. The is $1.4–1.6 \times 10^{-2}$ and the effective velocity, 2.55×10^{-2} to 2.25×10^{-1} m/d. Regional discharge occurs to the ocean, the consumptive use is very limited as xerophytic vegetation prevails, and a specific case involves springs. In the hydrochemical analysis, difficulties arose due to the anisotropy and heterogeneity, and to the level, density, and detail of the information available. SGI aquifers have higher salinity and sodium chloride type water. In the SGS, water in the Patagonia Formation has low TDS, mainly of sodium chloride and/or sodium sulphate type, and some of sodium bicarbonate type. Water in the Santa Cruz Formation has very low TDS, is of sodium chloride and/or sodium sulphate type, or sodium bicarbonate type.

Keywords Geohydrological system · Components · Hydrodynamics · Recharge · Circulation · Discharge · Hydrochemistry

This chapter is central in the development of this work and, in turn, it is based on the key considerations expressed so far. It begins with an analysis of the state of the art in hydrogeological research in the context of the basin and its relationship with its environment.

© The Author(s) 2017   41
M.A. Hernández et al., *Hydrogeology of a Large Oil-and-Gas Basin
in Central Patagonia*, SpringerBriefs in Latin American Studies,
DOI 10.1007/978-3-319-52328-6_4

4.1 Hydrogeological Research

Despite the significance of groundwater in this arid region, as it supports the productive socio-economic activities, there is very limited specific knowledge on the issue, unlike the case of the general geology and the one connected to the production of HCs and metal mining.

The most prominent features that have led to such a situation are, generally speaking, as follows:

(a) The first studies that attracted scientific research were within the fields of geology and palaentology, very attractive disciplines due to the natural characteristics of the region.

(b) Until 1907, when oil was discovered, the water demand was very low. It aimed at supplying sheep farming activities at a regional level and small, isolated agricultural settlements in the colonies of the lower Chubut River valley, the Andean sector of the city of Esquel, and the Sarmiento Valley; in all of these cases the only available resource was surface water. It was precisely due to the need to establish a port for wool trading that a drilling in search of water was undertaken in the then village of Comodoro Rivadavia, where HCs were found.

(c) Such a discovery had a major impact on the local water requirements, both to supply the needs of the oil exploration and exploitation stages, and to cover the demands of an exponentially growing population, and the related businesses and services.

(d) The lack of watercourses led to the prospecting of springs, the first groundwater source to be exploited, with the help of the experience of the settlers of Boer (Dutch) origin, who arrived in the region as a consequence of the First Anglo-Boer War in South Africa. Thus, at first, the springs in Manantiales Behr, Rosales, Campamento Escalante, and Cañadón Baumann were used, until 1965, when the first aqueduct from Musters Lake was built; in 1999, another aqueduct with a larger capacity was added to it. The catchment system provides 150,000 m^3/d to supply a population of 350,000 inhabitants (Comodoro Rivadavia, Caleta Olivia, and Rada Tilly), by means of an aqueduct extending over 224 km. The supply is at present insufficient and the aqueduct itself, poorly efficient.

(e) As regard the Southern Flank in the Eastern Sector, when the organized exploitation of oil began in 1933 and developed greatly as from 1945, the growing water needs led to the extraction by means of wells into aquifers of the Patagonia and Santa Cruz formations, in the vicinity of the city of Caleta Olivia (Cañadón Quintar and Cañadón Esther), supplying 200,000 m^3 per month (González Arzac et al. 1988, 1991; Hernández et al. 2008a).

The data obtained from the work carried out up to 1980–1990 was not very useful. This was caused by the lack of specialization of such data, due to the approach used, as well as to the drilling tools and techniques typical of oil production. Besides, the reduced area of the sites did not allow the regionalization of the data.

The Cátedra de Hidrogeología de la Universidad Nacional de la Patagonia San Juan Bosco [Chair of Hydrogeology of the Saint John Bosco National University of Patagonia, UNPSJB] (1982) carried out the first regional contribution of relevance, but it was not until the 1990s that studies and research became more frequent, with the contribution of specialized consultancy and other research centres; the contribution of the authors of this book corresponds to such a stage.

It is probable that the development in the exploitation of unconventional reserves will make it necessary to undertake a well-organized general assessment of the groundwater resources, as well as the environmental geohydrological analysis of the possible consequent negative impact (Hernández 2015).

In general, the absence of deep hydrological knowledge is a consequence of the lack of programmes on the subject funded by the government or by scientific organizations, as well as the result of the confidentiality of the information provided by private operators, such as those of the oil industry. The activity of scientific researchers, though valuable, lacks pragmatic or structural support and, frequently, enough economic resources.

4.2 Physical Component of the Local Geohydrological System

The first orderly approximation to aquifer behaviour in the San Jorge Gulf Basin can be found in studies by Grizinik (Cátedra de Hidrogeología UNPSJB 1982; Grizinik and Fronza 1996), referring to a multi-unit aquifer complex that gathers the formations with such a character underlying the Salamanca Formation.

In order to identify and define a geohydrological system, it is necessary—apart from locating the groundwater-bearing levels—to resort to other elements or units on the basis of their lithological behaviour as regard the water resources (Custodio and Llamas 2001). They are as follows:

Aquifers   (*aqua*: "water" + -*fer*: "bearing") Rocks or sediments that easily receive, contain, and transmit water (e.g., coarse-grained silts, sands, and gravels)

Aquicludes   (*claudere*: "to shut, close") Those which receive and bear water, but do not transmit it (i.e. clays)

Aquitards   (*tardare*: "to slow") Those which receive and bear water, but transmit with difficulty and under certain conditions (i.e. clayey silts and silty clays)

Aquifuges   (*fugere*: "to flee") Those which receive water, but neither bear nor transmit it, as in the case of hard rocks with secondary porosity (e.g., granites, diabases, diorites)

The physical component of a geohydrological system is, therefore, composed of the succession and interrelationship between the different hydrolithological units defined above, with all of the dynamic and hydrochemical events of the system occurring within it.

In the basin, two systems were identified; they had already been defined by Hernández et al. (2009b) and Hernández and Hernández (2013) as Upper Geohydrological System (Sistema Geohidrológico Superior, SGS) and Lower (or Passive) Geohydrological System (Sistema Geohidrológico Inferior o Pasivo, SGI).

The former system is active, that is to say, that it has a certain degree of connection with the present-day hydrological cycle, whose relevance is lost in depth. It comprises the Cenozoic lithostratigraphic units described in Chap. 3 and located in depth with respect to the Salamanca Formation, which it also includes (Table 3.1); this system occurs in a mainly porous medium.

The SGI is different, as it constitutes a mostly fissured medium and, in certain cases, with double porosity. It underlies the Salamanca Formation and reaches up to the volcano-sedimentary complex (Sylwan et al. 2011); it is not known whether the "hard" Lower Jurassic rocks actually behave as aquifuges or if they have secondary porosity, and—if this were the case—in which intervals of the unit they occur. The volcano-sedimentary complex is, to some extent, a productive fissured aquifer that supplies the metal mining activity to the south of the basin, at the expense of the Chon Aike and La Matilde formations (Hernández et al. 2011).

In Table 4.1, taking as a basis the stratigraphic scheme shown in Table 3.1, the different lithostratigraphic units are shown together with their hydrolithological counterpart, which—whenever there is more than one—are represented by different symbols. The width of the column corresponds to the approximate proportion with respect to the total geological thickness of the unit, which was observed by means of the geological cross-sections and/or electrical profiles used.

As regards the SGI, there is very little specific data, since the source of the information available is the oil industry and—as it is well known—the research methodology is very different. Besides, at the time, no importance had been given to this system, as the hydrogeological assessment is recent.

The bearing formations are generally HC reservoir rocks as well, whereas aquicludes are source rocks, simply due to their relative permeability (Schiuma et al. 2002).

In a fissured medium, the underlying Bahía Laura Group (Hernández 2000; Hernández et al. 2011) offers significant aquifer conditions that, as stated above, support an important gold mining activity, as well as being used for sheep watering in the rural area. Its secondary porosity is represented by diaclases in ignimbrites of the Chon Aike Formation and by open bedding planes in tuffs of the La Matilde Formation, both of which are components of the above-mentioned group. To the south of the Deseado River, the rate of flow is over 20–25 m^3/h.

In the basin, the deepest occurrence of aquifers in a porous medium coincides with the base of the Pozo Cerro Guadal Formation in the Yacimiento Río Mayo (personal communication). In other deep formations, such as the Pozo D-129 and Matasiete formations, there are only some thin, discontinuous, sometimes lenticular groundwater occurrences, which do not qualify as actual aquifers.

Table 4.1 Hydrogeological scheme

GROUP	FORMATION	HYDROGEOLOGICAL BEHAVIOUR	
	Santa Cruz Fm.	aq	
	Patagonia Fm.	aq	at/ac
	Sarmiento Fm.	at/ac	
	Río Chico Fm.	ac	at/ac
	Salamanca Fm.	ac	at/ac
CHUBUT	Bajo Barreal Fm. (upper member) = Yac. El Trebol Fm. = Yac. Mes. Espinosa Fm.	ac	at/ac
	Bajo Barreal Fm. (lower member) = Com. Rivadavia Fm. = Cañadón Seco Fm.	ac	
	Castillo Fm. = Mina del Carmen Fm.	ac	aq
	Matasiete Fm. + Pozo D-129 Fm.	ac	
NEOCOMIAN	Pozo Cerro Guadal Fm.	ac	at/ac
	Pozo Anticlinal Ag. Bandera Fm.	ac	
BAHÍA LAURA	La Matilde Fm.	af	
	Chon Aike Fm.		

aq aquifer, *at* aquitard, *ac* aquiclude, *af* aquifuge

It is only the Castillo (= Mina del Carmen) Formation that acts as an aquifer in fluvial and lacustrine sands; no further information is available, except for the occurrence of several thin layers with high salinity. In the Comodoro Rivadavia and Cañadón Seco formations (lower member of the Bajo Barreal Formation) only the sporadic presence of groundwater has been observed, occurring discontinuously, judging by the considerable differences in closely located wells (Hernández 2015).

The El Trébol Formation and its equivalents (Yacimiento Meseta Espinosa and Laguna Palacios formations) are aquifers in their basal sector. The formation water that accompanies the exploitation of HCs is saline, with moderate discharges, although in certain cases these are greater than those of other oil units. It is usually reused after injection operations in the secondary recovery of oil.

The SGS is the most important system for the objectives of this work due to its relationship with the present-day hydrological cycle, as mentioned above.

Beginning with the Salamanca Formation, has a poor *aquifer* capacity at its base, composed of glauconite sandstones where a water-oil-gas system occurs and in which the original finding of HCs took place. As shall be explained below, the confined aquifer bears saline water, except in one unit in the Western Sector (un-documented) in which fresh to brackish water was found. The upper portion is an *aquiclude*, with plastic marine clays.

In the Río Chico Formation (or Río Chico Group), there occur different lithologies, from bentonite clays and very fine-grained tuffs to tuffaceous sand-stones and well-cemented conglomerates, but in general they coincide in their very low permeability. This makes it possible to describe the Río Chico Formation as an *aquiclude*, with the occasional occurrence of water-bearing sandstones, but without the continuity that would allow its accurate definition as an aquifer.

The Sarmiento Formation is an *aquiclude* due to the dominance of fine-grained pyroclastic rocks (tuffs and ash tuffs) and to the absence of water-bearing levels in the oil exploration/exploitation wells. Besides, it is generally regarded as the lower impermeable layer or hydrogeological base of the Patagonia Formation (Cátedra de Hidrogeología UNPSJB 1982; Grizinik and Fronza 1996; Hernández et al. 2008b; Hernández and Hernández 2013).

It is precisely the Patagonia Formation—together with the overlying Santa Cruz Formation—that is the most important aquifer unit in the basin (generically grouped as Patagonian aquifer), as a result of its extension, yield, and diversity of uses. It is predominantly located in the Eastern Sector, with its useful thickness decreasing towards the west, and it may exceed 100 m.

Certain distinctive features are worth mentioning; for instance, not all of the thickness of the formation is an *aquifer*, as in many wells it is entirely clayey and, therefore, an *aquiclude* (Hernández and Hernández 2013). It is necessary to clarify that the presence of this lithostratigraphic unit in a certain area does not imply that it acts as an aquifer, which leads to conflicts between the oil companies and the authorities enforcing the regulations in the provinces of Chubut and Santa Cruz. This situation is mainly observed in the Southern Flank of the Eastern Sector of the basin and also to the southwest, where there are intermediate cases with sections with different behaviours in the same borehole. In order to assess the different hydrolithological behaviours of the Patagonia Formation, three patterns identified by the site, where the reference well logs were obtained have been proposed: the Cañadón León pattern (complete aquifer), Piedra Clavada pattern (partial aquifer), and Koluel Kayke pattern (complete aquiclude), following Hernández and Hernández (2013), as shown in Fig. 4.1.

Depending on its spatial location, when the Patagonian aquifer crops out, it may behave as a phreatic aquifer, especially in the Eastern Sector, and even share such a character with the so-called "Rodados Patagónicos" that overlie it, as a single hydraulic unit (Hernández et al. 2009b). It also occurs as semi-confined or confined, when the sandy strata underlie other pelitic strata, well-cemented sandstones, or coquinas in the same formation.

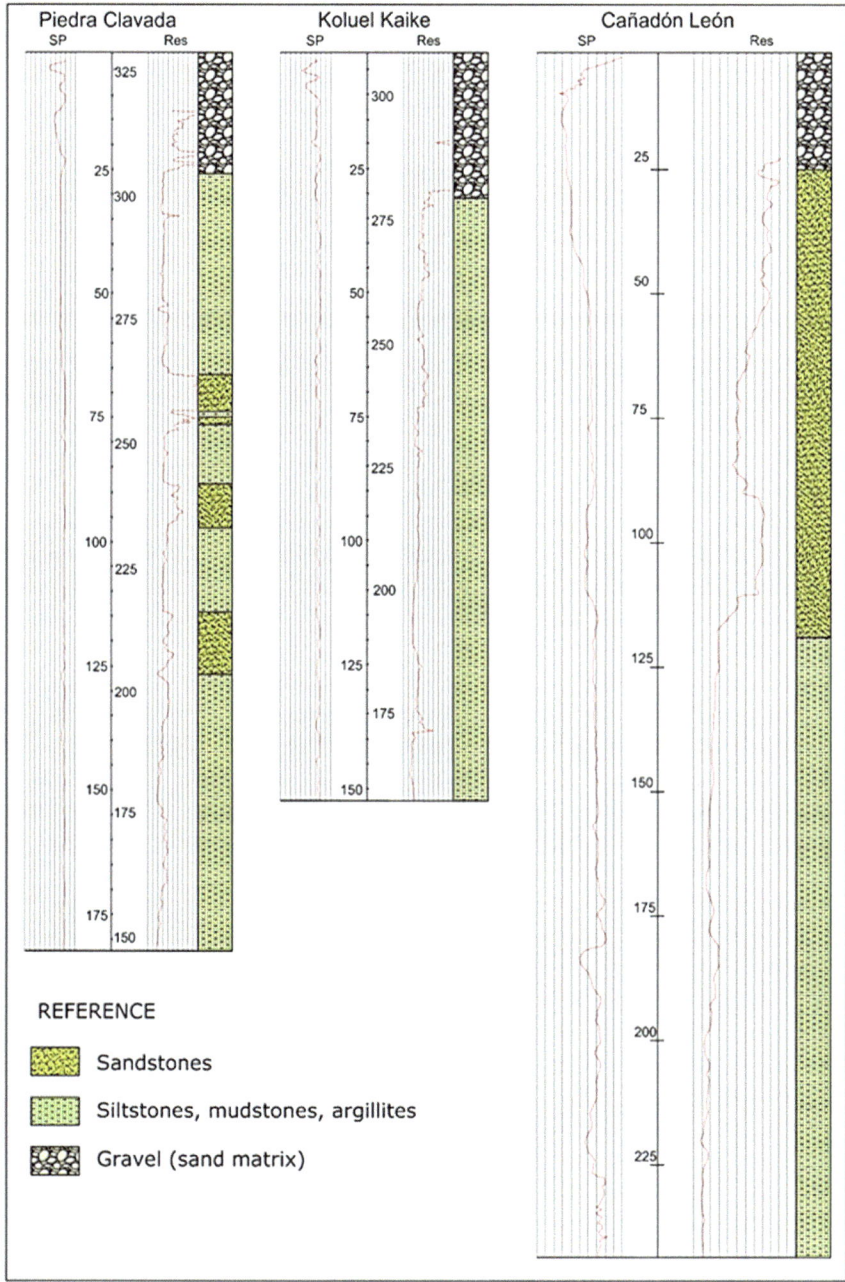

Fig. 4.1 Different patterns of hydrolithological behaviour in the Patagonia Formation, based on representative well logs

In the case of this aquifer, it is significant due to its marked anisotropy (horizontal and vertical) and at times its strong heterogeneity, as shown by the hydrolithological patterns of behaviour described above. This characteristic affects the regularity in the behaviour of the groundwater dynamics, but even more markedly the hydrochemical quality, with major areal changes that have made it difficult, for instance, to apply the ion contents to the entire region, as discussed in the corresponding section.

The general hydrochemical characteristics range from brackish through saline to practically fresh water, despite having a marine origin. When this occurs, it is usually confused with the aquifer of the Santa Cruz Formation, as shall be analyzed in Sect. 4.4.

As regard the more frequent yield values, they are of the order of 8–10 to 15 m^3/h (González Arzac et al. 1991; Plusagua 2011); it should be mentioned that there is more information available, though it is protected by the confidentiality imposed by oil companies. The average specific yields are between 100 and 150 m^3/d m, with transmissivity coefficients of over 80 m^2/d.

It has been, and still is, used for water supply to urban centres (Caleta Olivia, El Tordillo and El Trébol, for example), water injection for the secondary recovery of oil, small industries and even peri-urban horticulture.

In the Santa Cruz Formation occurs the aquifer of the same name, generally semi-confined to confined (except when it crops out or is covered by gravels), with a location closely related to the one of the Patagonia Formation, known as Patagonian aquifer. Until complete lithological descriptions—not limited to the sections of interest to the oil industry—and multi-profile geophysical records started being made in recent years, the criteria to differentiate these two formations depended on the presence of marine fossils in the injection returns or when outcrops of interpretive support occurred. In any case, there are still difficulties with the information provided by most of the prior productive drillings.

As the Santa Cruz Formation overlies the Patagonia Formation, it may have been the case that the infiltration processes progressively dissolved the chlorides in the aquifer occurring in the latter and decreased its total dissolved solids (TDS). On the other hand, the exploitation yields and hydraulic parameters are also very similar. Besides, from the legal point of view, the legislation of the Province of Santa Cruz does not differentiate between them (Hernández and Hernández 2013), as shall be analyzed below.

At the top of the hydrogeological sequence are the Pleistocene "Rodados Patagónicos" that are part of the vadose zone or host the phreatic aquifer; in the former case, when they are at a shallow depth. The vadose zone may sometimes reach an important development of over 20 m in the tableland areas (Hernández et al. 2009a), while being shallow along the marine coastline.

In turn, the phreatic aquifer may also occur in a more permeable section of the underlying Tertiary deposits, or occur in them when the gravel cover is lacking, although this is not frequent. It is only used in the rural area, mainly for sheep watering.

4.3 Hydrodynamics

The mobile component of the system is analyzed following the hydrodynamic cycle, which comprises the *recharge, circulation* and *discharge* stages, from a comprehensive perspective and giving specific details when the level of information available allows it.

As regard the *recharge* of the system, it is necessary to revisit the concepts already discussed in Chap. 2 and Sect. 3.1 concerning the unsuitability of applying methods such as the one of Thornthwaite and Mather to calculate infiltration in arid regions.

In order to do so, considering the actual occurrence of groundwater in the region, even with low salinity, the mechanisms defined by Hernández (Hernández 2000; Hernández et al. 2008b, 2009a) for arid zones were identified in the first place, so as to subsequently estimate the possible values by resorting to the experience gathered in neighbouring regions.

The mechanisms analyzed specifically in the Extra-Andean Patagonia are the actual reduction in consumptive losses, rapid infiltration, rapid concentration, delayed recharge and the influence of losing streams, as well as the coincidence of two or more of them.

The *actual reduction in consumptive losses* (Fig. 4.2) is caused by the dominant occurrence of xerophytic vegetation sensu Paruelo et al. (1992) (Figs. 4.3 and 4.4), with a high degree of specialization for water retention and features, such as aphylly, lack of stomata, succulence, ephemeral flowering, coriaceous epidermis and others.

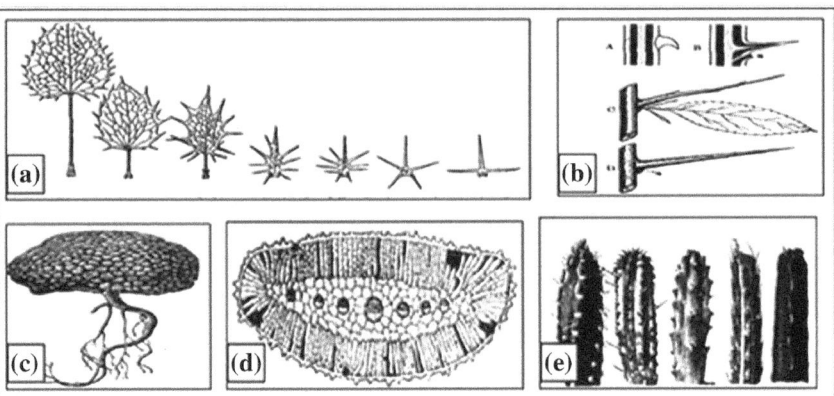

Fig. 4.2 Frequent examples of local xerophytic specializations; **a** evolution from a mesophytic leaf to a xerophytic thorn; **b** from an axillary bud to a prickle or thorn; **c** a case of coriaceous epidermis (*Azorella compacta* or yareta), very frequent in the south of the basin; **d** palisade parenchyma of *Tamarix gallica*, a shrub of Eurasian origin introduced in Chubut and Santa Cruz in the 19th century; and **e** succulence in some examples of cacti species

Fig. 4.3 Xerophytic herbaceous vegetation with low density cover on the Pampa del Castillo tableland, to the south of Colhué Huapi Lake. Courtesy of Julio Cotti Alegre

In practice, such specializations lead to a lack of evapotranspiration—as in the case of mesophytes—and for part of the rainfall contribution not to be subtracted, to the benefit of infiltration (Hernández et al. 2008b).

Fig. 4.4 Low shrubland vegetation in a gully in the Cerro Dragón area. Courtesy of Julio Cotti Alegre

Rapid infiltration, which is a fairly common process, occurs for two main reasons: the usually high permeability of porous materials in arid zones (Fig. 4.4) or the occurrence of fissures in rock materials, in both cases when the soil specific retention is low to null (Hernández et al. 2002). This leads to the early subtraction of the net input to the consumptive use, a process aided by the usually deep water table, which is beyond the physical possibility of evaporation or the penetration of phreatophytic plant roots (Lerner et al. 1990; Hernández 2000). Aridisols, Entisols or Inceptisols have low specific retention and, as is generally known, infiltration occurs once the field capacity of the soil has been saturated. Values below 10 mm (Ferrer et al. 1990) make it easier for rapid infiltration to occur, when the permeability in the vadose zone is high.

Rapid concentration occurs in the periphery of hills or tableland edge areas, adjacent to permeable deposits in low relief zones (Senguerr River valley, lake area, Deseado River valley). It is significant when the slope is steep and the substrate is not very permeable, facilitating infiltration in the lower parts (Fig. 4.5).

As regards *delayed recharge*, it is caused by the permanence of snow, frost and/or permafrost during winter, and their subsequent more or less rapid incorporation into the groundwater system at the beginning of the spring (Fig. 4.6). Therefore, the input after a few days is equivalent to the contributions of several months. Such permanence coincides with the occurrence of the minimum relative

Fig. 4.5 View of a typical landscape for rapid concentration occurrence. Rìo Senguerr basin. Courtesy of Hidroar SA

humidity values, maximum barometric pressure values and calms. It occurs in the main tablelands (Pampa de Salamanca, Pampa del Castillo, Meseta Espinosa, Pampa de los Guanacos, Meseta Pelada and others).

Fig. 4.6 Delayed recharge scenario. Sector to the north of the great lakes (Musters and Colhué Huapi). Photo by Mario Hernández

The *influence of losing streams* in the basin follows the regional pattern of the Extra-Andean Patagonia: they are allochthonous, with Andean headwaters, generally perennial and influent, and therefore they constitute lineal recharge areas (Alvarez et al. 2006). The oil industry exploits the water of the underflow of the Senguerr and Deseado rivers due to its high quality; the latter is the only intermittent stream in the region. The water of the Senguerr River underflow was used until 2016 for the secondary recovery of oil, with a rate of flow of the order of 10,000 m³/d.

It is usual for two or more of the mechanisms described above to occur concurrently, for instance, delayed recharge in tablelands with xerophytic vegetation, covered with very permeable gravels, or other possible combinations.

Recharge quantification is not easy for several reasons, among them the inadequacy of the hydrometeorological data as regards its density, quality and record, the lack of snow measurements in the tablelands (there are only records in the Andes), and of criteria in the allocation of areas of rainfall influence in the calculations.

In this case, due to the lack of time series rainfall data, the methodology introduced in Chap. 2 was applied: the extrapolation to the data of Comodoro Rivadavia of the coefficients derived from two methods implemented in two sites located in the vicinity of the Extra-Andean Patagonia, that is, Puerto Madryn (Álvarez 2010; Álvarez et al. 2013, 2016), located in a coastal area of the Atlantic similar to Comodoro Rivadavia, and Gobernador Gregores (Hernández 2000), in the inland area to the southwest of the basin.

In the first case, precipitations higher than 5 mm were selected and the Balshort daily time-step water balance (Carrica 1993) was used, and the results were compared with chloride balances and the mathematical modelling of the recharge. In the second case, selected precipitations were also used, as well as a microbalance using the measurement of discharge in a spring system as control.

For a 175-mm selected rainfall, infiltrable excess values between 59 and 76 mm were obtained, representing 30 and 33.5% of the gross rainfall, and 33.4 and 43.4% of the selected rainfall, respectively.

These values are consistent with the discharge circulating through the SGS and with the volumes historically withdrawn for public supply in Comodoro Rivadavia and by the oil industry in the basin.

There is no equivalent information available on level and density so as to be able to reconstruct the *groundwater circulation* in the SGI and SGS. In the first case, it is insufficient mainly because the data was obtained from wells built for the use of the oil industry, with a design that is inconvenient even for the measurement of water level depth, lacking in hydraulic tests to determine specific parameters and with no reliable differentiation of the aquifers.

In the SGS, the data are better in the Patagonian and Santa Cruz aquifers; at least, it was possible to construct an equipotential diagram, which is shown in Fig. 4.7, as a synthesis of a more detailed map which may not be reproduced due to the size constraints of the book.

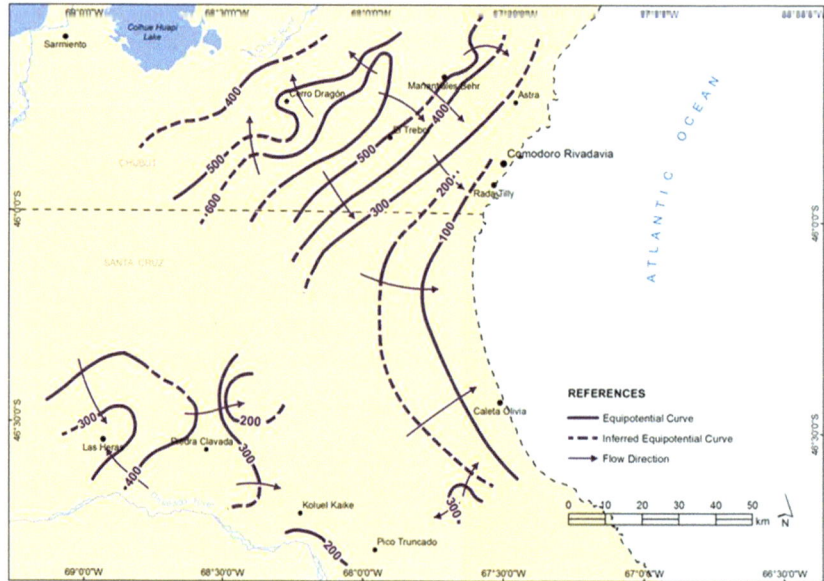

Fig. 4.7 Equipotential diagram

Not enough information is available on water levels due to the low density and the doubtful reliability of the data, which makes it impossible to reconstruct the *groundwater circulation* in the SGI. The information on the SGS is more suitable, although it is not ideal, as it shares certain defects with the one mentioned above, as discussed in Chap. 2 when addressing the methodology used.

An equipotential diagram was constructed with the potentiometric surface of the Patagonian aquifer, without discriminating between the Patagonia and Santa Cruz formations; a broken line was used when showing an inferred section of a curve (Fig. 4.7). The original scale used was 1:5000, but it was then converted to another one which would be more compatible with the book format. The equidistance chosen is related to the scale, to the possible level of detail, and the schematic character intended. Flow directions were marked perpendicular to the contour lines, whose morphology is of a radial type, with a certain trend towards a planar type in some areas, especially in the eastern flank of the tablelands.

It should be taken into consideration that the source formations tend to disappear towards the west, as mentioned in Sects. 3.4 and 4.2, which is why the diagram does not extend in that direction. It does not include the central area either, as— together with such a disappearance—there are usually no oil reservoirs, and if there are, they lack wells that would provide suitable information.

A general direction of the regional flow towards the east (Atlantic Ocean) may be observed, with divergent sectors such as the one located to the southeast of the Cerro Dragón, where the runoff can be locally recognized from a SW–NE-trending groundwater divide towards the large depression where the Musters and Colhué Huapi lakes occur. Another local flow situation is shown in the bottom left of the diagram, centred in the region around the city of Las Heras.

In general, the diagram sketched shows certain regularity in the scale and equidistance used, with average hydraulic gradients of the order of 1.4×10^{-2} to 1.6×10^{-2}. The average minimum values are of approximately 8×10^{-3} to 1.1×10^{-2} and the maximum ones reach from 1.5×10^{-1} to 4×10^{-1}.

As regard the estimation of the effective flow velocity, it was based on the following equation:

$$V_e = Ki/\emptyset, \tag{4.1}$$

where V_e is the effective velocity (m/d), K is the permeability coefficient (m/d), i is the hydraulic gradient (−), and $\emptyset$ is the effective porosity coefficient (−).

The K values used were cited from previous contributions (Cátedra de Hidrogeología UNPSJB 1982; Hernández et al. 2008b; Auge et al. 2009) and provided by Consultora GM&A. In general, the most reliable ones are between 0.17 and 1.5 m/d. The value of $\emptyset$ would be 0.10, according to the lithology of the formation and to the data mentioned by Hernández et al. (2008a); therefore, using the average hydraulic gradient (1.55×10^{-2}), the effective velocity would be of the order of 2.555×10^{-2} to 2.255×10^{-1} m/d. If the maximum gradient of 45×10^{-1} were applied, a V_e of 6×10^{-1} m/d could be reached.

In the aquifers of the SGS, such low effective flow velocities, which entail a prolonged time of residence of water in the aquifers, together with the deficient hydroclimatic conditions, explain the hydrochemical properties discussed below, especially the greater occurrence of brackish and saline waters.

The regional *discharge* that predominates in the system is to the Atlantic Ocean, at different distances from the coastline, depending on the formation concerned, its offshore penetration, and the difference in charge with respect to the height of entry into the subsurface. The consumptive use, even though the value of the potential evapotranspiration is very high in relation to the pluvial and pluvio-nival contributions, is restricted by the type of xerophytic vegetation, specialized in the conservation of the scarce water available, forced by the arid climate of the region. Locally, the equipotential diagram shows other areas of local discharge, morphologically depressed and with centripetal flow.

Springs are areas of local discharge and their historical relevance in the region has already been discussed. Most of them are of a *stratigraphic* type and occasionally *structural* or of a *talweg* type, these last two are phreatic and the first one generally originates in the Patagonia and/or Santa Cruz formations. Discharging very high-quality water, they are usually located in low positions of the relief or on the sides of gullies, generally supplying the wetlands locally known as *mallines*, rural populations, and even localities of a certain importance, at least partially. As the watercourses are losing or influent streams, they are not involved as a discharge phenomenon (Alvarez et al. 2006).

The influence of the present-day anthropogenic discharge is of minor relevance at a regional scale, as Comodoro Rivadavia, Rada Tilly and Caleta Olivia have been supplied for many years with surface water from the Musters Lake, as mentioned above, whose contributions replaced the local groundwater sources. As regard the oil industry, in this territory, to the south of the parallel 46°S, within the jurisdiction of the Province of Santa Cruz, the regulations limiting its use are applied. The progressive increase in water/oil ratio, which leads to the reuse of water for the secondary recovery of oil and gas, also has an impact.

It should be mentioned in this chapter that, despite its schematic character (Fig. 4.7), the equipotential diagram constructed as of the date of publication of this book is the only regional attempt at doing so in such a vast area of the Extra-Andean Patagonia.

4.4   Hydrochemistry

As discussed above, there were difficulties when dealing with this chapter, arising from the anisotropy and heterogeneity of the aquifers, and from the poor level, density, and detail of the information available.

Regarding the anisotropy, its main impact was on the impossibility of regionalizing the analytical results. Therefore, the aquifers of the SGS in the Santa Cruz and Patagonia formations had to be grouped into diagrams, such as the Piper and

Schoeller-Berkaloff plots. On the other hand, the aquifers of the SGI remained undifferentiated and they mainly occur in the El Trébol Formation, as well as in the Castillo and Mina del Carmen formations, being aimed at water injection for secondary recovery (Hernández 2015).

Concerning the information available, it was mostly from oil wells until recent years, when monitoring wells started being set up; they were specifically built with hydrogeological criteria and with adequate geometry, design, and materials for the performance of hydraulic tests. There is a certain degree of uncertainty regarding the representativeness of the samples obtained from the oldest wells; to a large extent such data was replaced by supplementary information and the knowledge of the deposits. The samples selected for treatment are characteristic of the area of the basin with reliable data (documented wells and original analytical protocols), having avoided subjective bias.

In general, the aquifers of the SGI have higher salinity—with TDS values from 5000 to 20,000 mg/L, corresponding to brackish and saline waters—and hardness, whose most frequent values are of the order of 1600–6000 mg/L as CO_3Ca.

On the other hand, in the SGS, the TDS is very low in the groundwater of the Santa Cruz Formation, from 288 to 434 mg/L and only one sample of 1130 mg/L, and higher in the samples of the Patagonia Formation, from 530 mg/L to slightly over 3000 mg/L, with a single sample of 182 mg/L. This exceptional sample could be an example of the difficulty in differentiating both formations in certain circumstances, as discussed below.

For the treatment of ion contents, Piper and Schoeller-Berkaloff plots, already described in Chap. 2, were used. The water samples from the SGI were grouped and the components of the Patagonian aquifer (Patagonia and Santa Cruz formations) were treated separately; there is more information on the Santa Cruz Formation, although it is difficult to differentiate both formations in practice. Dissimilar anionic behaviours can be observed in the Patagonia Formation.

First, regarding the plots for the deep aquifers of the SGI (Fig. 4.8), in the Piper plot, the predominance of the sodium chloride-sulphate facies—basically, a sodium chloride type water and secondarily a sodium sulphate type—can be noticed.

Among the anions, the presence of bicarbonates is negligible, whereas as far as cations are concerned, calcium follows sodium in order of abundance.

In the Schoeller-Berkaloff diagram, it can be observed that, besides confirming the facies grouping discussed above, these are mature or evolved waters, with long time of residence and consequent contact with the lithological material, probably more due to storage permanence than transit time. In the chloride water samples, this anion is strongly dominant over bicarbonate and sulphate, in that order. In the sulphate water samples, chloride follows in order of abundance, exceeding bicarbonate.

The representation of the samples selected from the SGS makes it possible to observe a difference in behaviour between the aquifers of the Patagonia and Santa Cruz formations, as well as of both of these with respect to those of the SGI (Grizinik et al. 2003). Starting by the interpretation of the samples from the Patagonia Formation, in the Piper plot (Fig. 4.9) most of them may be placed within

Fig. 4.8 Piper and
Schoeller-Berkaloff plots. SGI
aquifers

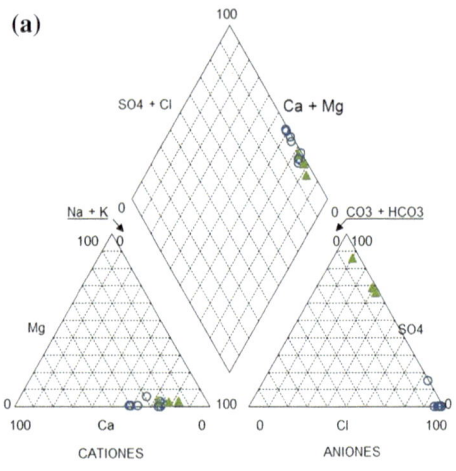

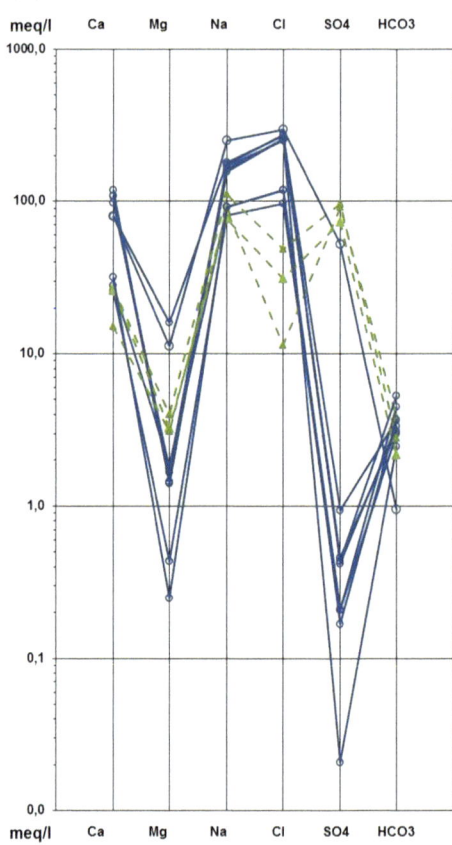

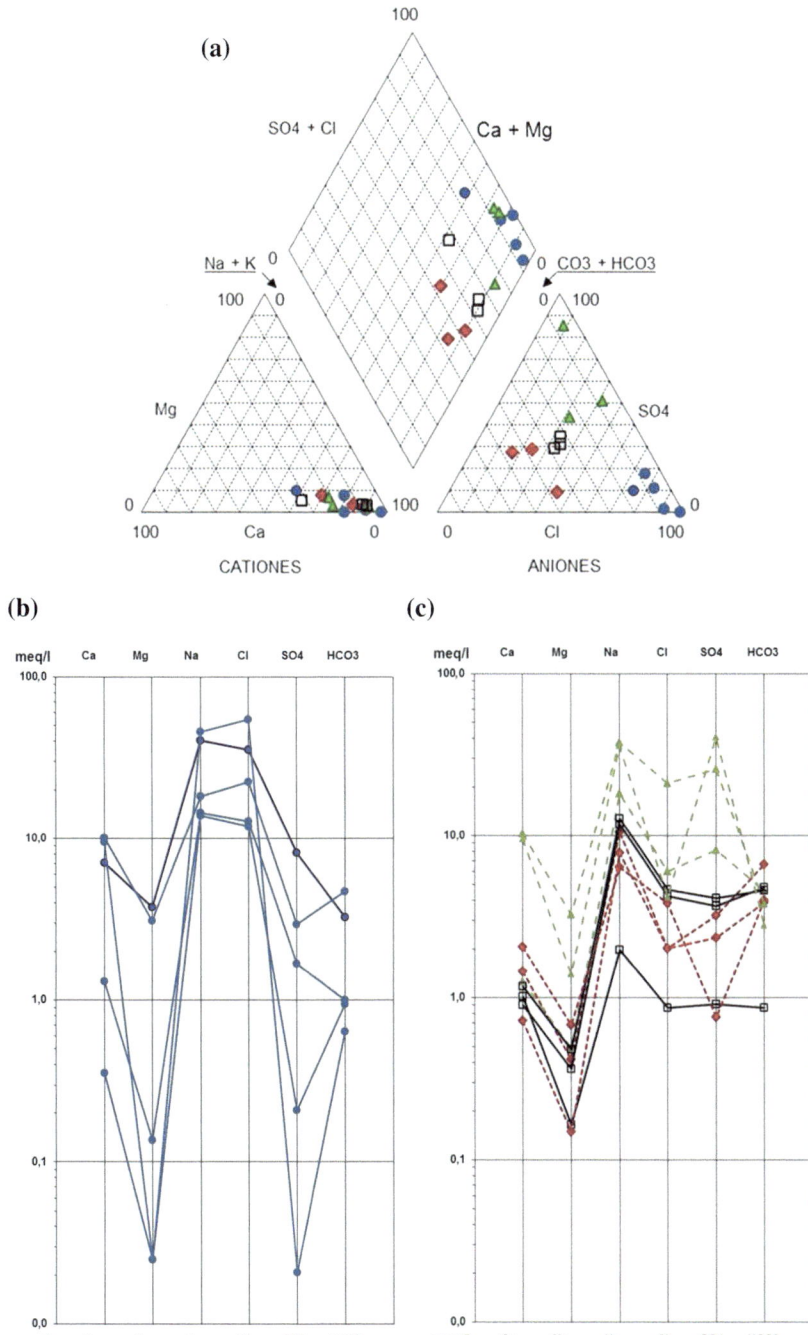

Fig. 4.9 Piper and Schoeller-Berkaloff plots. SGS aquifers of the Patagonia Formation

the domain of sodium chloride and/or sodium sulphate facies, and some of them within the one of sodium bicarbonate. In the first facies, in turn, the chloride type water prevails over the sulphate type, and over others in which there is no dominant ion. Chloride is distinctly dominant, with over 75% of the anion total, in a chloride-sulphate-bicarbonate sequence, and in all cases with a clear sodium character. The sulphate type has chloride as a subordinate anion, whereas the bicarbonate type has sulphate. As can be observed, it is a true anionic dispersion, unlike the case of cations, in which sodium is always the most abundant.

In Fig. 4.9, the Schoeller-Berkaloff diagram shows the above-mentioned characteristics. It was decided to use two diagrams to facilitate observation: one of them showing the chloride waters, which are the most abundant, and the other including the sulphate and bicarbonate waters, as well as those in which neither of the anions exceeds 50%. In the first diagram, those with a more mature or more evolved character can be observed, whereas in the second diagram, bicarbonate waters have a younger configuration or reflect more recent recharge. These two plots confirm the above-mentioned ion dispersion.

The Piper and Schoeller-Berkaloff diagrams constructed to display the analytical results of the aquifer of the Santa Cruz Formation are shown in Fig. 4.10. Two facies domains can be distinguished in the Piper plot: one of the chloride type water and the other of the sodium bicarbonate type. The first type is the one with higher TDS values (494–1130 mg/L), showing a higher chloride content, and with more bicarbonate than sulphate. The second one, in turn, shows a noticeable prevalence of bicarbonate over chloride and sulphate, in that order; this prevalence is less marked in the case of cations in which sodium, even though dominant, has an uneven abundance with respect to the alkaline-earth type.

In the Schoeller-Berkaloff diagrams, a high degree of similarity can be observed in the broken lines for the different samples, suggesting greater hydrochemical homogeneity in the group. This feature is the result of several factors, such as the greater external influence due to its position in the geological sequence, the less evident anisotropy than in the other aquifers, and the relative regional homogeneity.

As a general comment, it could be pointed out that both the aquifer of the Santa Cruz Formation and those of the SGI have a narrower threshold of facies variations, and even TDS with smaller orders of magnitude, with clearer ion groupings. On the other hand, in the Patagonia Formation, the dispersion of the analytical values is the most noticeable characteristic.

A hydrochemical behaviour that is also common in other sectors of the Extra-Andean Patagonia is the role of sulphates due to the frequent presence of gypsum in the lithological package. This feature indicates the importance of the geological imprint with respect to the correlation of the hydrodynamics with the hydrochemical evolution (Consejo Federal de Inversiones 1986; González Arzac et al. 1991; Hernández and Hernández 2013).

It is extremely difficult to be able to specify the degree of influence of the different factors concerning the geohydrochemistry, from the original factors to the phenomena modifying the composition. The influence of the regional climate could be of relevance only in the SGS, since the deeper aquifers are very distant in time and space

Fig. 4.10 Piper and
Schoeller-Berkaloff plots.
SGS aquifers of the Santa
Cruz Formation

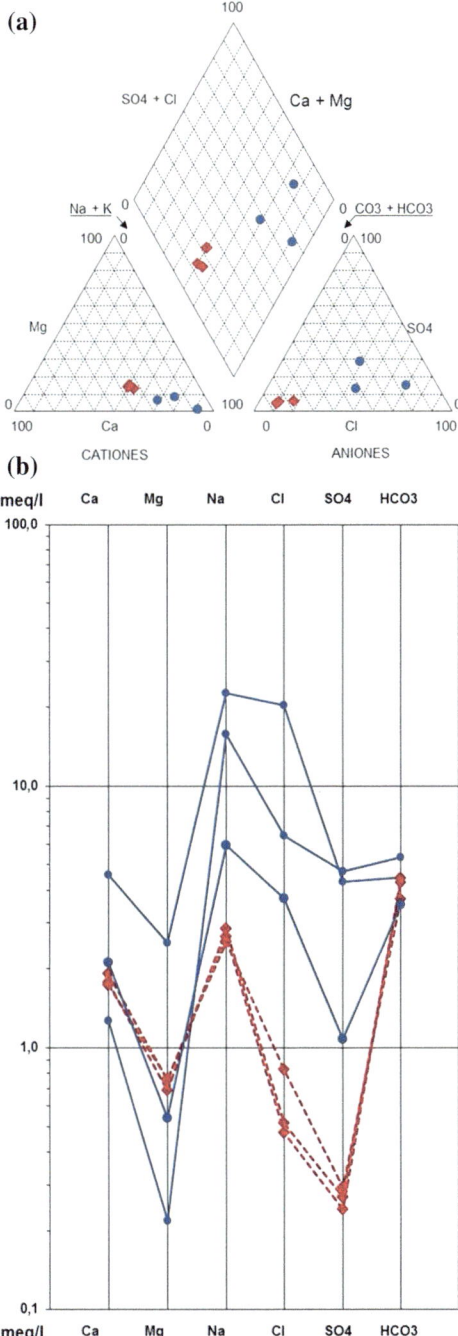

from the external events. The chemical evolution due to circulation could not be verified by means of the comparison of the analytical information with the equipotential map of the aquifers of the SGS (Fig. 4.7) and, as explained above, there is insufficient information on the lower ones. As a result of the lack of monitoring wells in the region, the same inconvenience arises when trying to assess the effects of the Senguerr River, which dilutes the concentrations of the upper aquifer levels.

Therefore, until new reliable data is generated, the lithological condition remains as the most outstanding factor, such as the above-mentioned presence of gypsum in the formations with water-bearing sulphates in solution, where the phenomenon of the dissolution of gypsum or anhydrite may occur. Other influences that may be relevant are the anisotropy recognized and, at times, the heterogeneity in the regional extent, which can be verified at least in the Patagonia and Santa Cruz formations (Hernández and Hernández 2013). Such influences may be responsible for the hydrochemical changes over short distances that occur in the oil exploitation area with the highest density of data.

As regards the minor components, proportions of HCs (TPH, BTEX aromatics, PAHs) have been found, reported by the enforcement authority of the provinces (personal communication) in different aquifer levels in the areas of oil and gas production. Occasionally, heavy metals and nitrates have been observed, as well as others of natural genesis, such as boron and strontium.

References

Álvarez MP (2010) Investigación geohidrológica de un sector de Peninsula Valdes, provincia de Chubut. Unpublished doctoral thesis, Facultad de Ciencias Naturales y Museo, Universidad Nacional de La Plata, 338 pp

Álvarez MP, Hernández L, Hernández MA, González N (2006) Relación aguas subterráneas-aguas superficiales en Patagonia Extrandina. República Argentina. Revista Latino-Americana de Hidrogeología. 6:43–48 (Montevideo, Uruguay)

Álvarez MP, Trovatto MM, Hernández MA, González N (2013) Estimación de recarga en zonas áridas según distintos métodos. Área medanosa del sur de Península Valdes (provincia de Chubut). In: González N, Kruse EE, Trovatto MM, Laurencena P (eds) Agua subterránea recurso estratégico. EDULP, La Plata, vol I, pp 46–51

Álvarez MP, Trovatto MM, González N, Dapeña C, Bouza, PJ, Hernández MA (2016) Estimación de la recarga en la zona medanosa de Península Valdés y su tendencia. In: García RF, Rocha Fasola V, Dornes P (eds) Relación Agua superficial-agua subterránea. Editorial Científica Universitaria, Catamarca, pp 229–236

Auge M, Simeone A, Rodríguez JJ (2009) Hidrogeología de Manantiales Behr Comodoro Rivadavia-Chubut-Argentina. In: Mariño EE, Schulz CJ (eds) Aportes de la Hidrogeología al conocimiento de los Recursos Hídricos. AIHGA-Amerindia, Santa Rosa, vol I, pp 85–95

Carrica JC (1993) El Balshort. Un programa de balance hidrológico diario del suelo aplicado a la región sudoccidental pampeana. XII Congreso Geológico Argentino, Actas VI:243–248 (Mendoza)

Cátedra de Hidrogeología UNPSJB (1982) Reconocimiento Geohidrológico del Sudeste de la Pcia. del Chubut. Universidad Nacional de la Patagonia San Juan Bosco (UNPSJB) – Dirección general de Recursos Hídricos (DGRH). Unpublished, Comodoro Rivadavia

Consejo Federal de Inversiones (1986) Geología y Geomorfología del NE de la Provincia de Santa Cruz. Consejo Federal de Inversiones Tomos I y II, Buenos Aires 89 pp

Custodio E, Llamas MR (2001) Hidrología Subterránea, Tomo I and II, 2nd edn. Omega, Barcelona 2350 pp

Ferrer J, Irisarri JA, Mendia M (1990) Cartografía de los suelos. In: Estudio Regional de Suelos de la Provincia del Neuquén. Consejo Federal de Inversiones (CFI)—Consejo de planificación y acción para el desarrollo (COPADE) – Provincia del Neuquén. Buenos Aires, pp 1–3

González Arzac R, Barbagallo J, Pérez Spina R (1988) Evaluación del sistema de explotación Cañadón Quintar. Caleta Olivia. Informe Final. Consejo Federal de Inversiones, 67 p, Buenos Aires, 50 pp

González Arzac R, Díaz JL, Calvetty Amboni B (1991) Geohidrología del Área Noreste de la Provincia de Santa Cruz. Consejo Federal de Inversiones. Serie Investigaciones Aplicadas, Colección Hidrología Subterránea N° 1, Buenos Aires 39 pp

Grizinik M, Fronza S (1996) Hidrogeología de la región situada al Noreste de Las Heras, Provincia de Santa Cruz, Argentina. XIII Congreso Geológico Argentino y III Congreso de Exploración de Hidrocarburos, Actas IV:417–425 (Buenos Aires)

Grizinik M, Pezzuchi E, Locci F (2003) Caracterización hidroquímica de las aguas subterráneas del Centro Norte de la Pcia. de Santa Cruz. I Seminario Latinoamericano sobre Temas Actuales de la Hidrología Subterránea, Memorias 2:451–459, Rosario

Hernández MA (2000) Estudio geohidrológico de la región Cerro Rubio-Cerro Vanguardia, provincia de Santa Cruz. Unpublished doctoral thesis, Facultad de Ciencias Naturales y Museo, Universidad Nacional de La Plata, 163 pp

Hernández MA (2015) Recursos hídricos. In: Recursos hidrocarburíferos No Convencionales shale y el desarrollo energético de la Argentina. Caracterización, oportunidades, desafío. EUDEBA, Buenos Aires, vol 4, pp 307–348

Hernández L, Hernández MA (2013) Características hidrolitológicas de las formaciones Patagonia y Santa Cruz. Cuenca del Golfo San Jorge. (Provincias de Chubut y Santa Cruz). In: González N, Kruse EE, Trovatto MM, Laurencena P (eds) Agua subterránea recurso estratégico. EDULP, La Plata, vol I, pp 112–117

Hernández MA, González N, Sánchez RA (2002) Mecanismos de recarga de acuíferos en regiones áridas. Cuenca del Río Seco, Provincia de Santa Cruz. Argentina. XXXII IAH International Congress—VI Congreso ALHSUD, Mar del Plata. doi:10.1007/s00254-004-0989-0

Hernández L, Hernández MA, González N, Ceci JH, Sánchez R (2008a) Origen de aguas subterráneas salinas en la zona de Caleta Olivia. Provincia de Santa Cruz. Argentina. IX Congreso Latinoamericano de Hidrología Subterránea, ALHSUD. Quito [CD-ROM]

Hernández MA, González N, Hernández L (2008b) Late Cenozoic geohydrology of Extra-Andean Patagonia Argentina. In: Rabassa J (ed) The Late Cenozoic of Patagonia and Tierra del Fuego, Developments in Quaternary Science, Elsevier, Amsterdam, vol 11, pp 497–509

Hernández MA, González N, Hernández L (2009a) Regiones áridas. Procesos diferenciales de recarga y casos ejemplo de Argentina. In: Carrica J, Hernández MA, Mariño E (eds) Recarga de acuíferos. Aspectos generales y particularidades en regiones áridas. AIHGA-Amerindia, Santa Rosa, pp 63–70

Hernández MA, Scatizza C, Rojo M, Preiato S, Hernández L (2009b) Un método para estimar la sensibilidad hidrológica aplicado en la cuenca del Golfo San Jorge. Provincias de Chubut y Santa Cruz. Boletín Geológico y Minero de España 120(4):523–532 (Madrid)

Hernández L, Hernández MA, González N (2011) Modelos de sistemas geohidrológicos en Patagonia Extrandina. In: García RF, Rocha Fasola MV (eds) Hidrogeología Regional y exploración hidrogeológica. Asociación Civil Grupo Argentino-AIH, Buenos Aires, pp 19–26

Lerner DN, Issar AS, Simmers I (1990) Groundwater recharge. International Association of Hydrogeologists (IAH), vol 8. Verlag Heinz Heise, Hannover, p 345

Paruelo JM, Aguiar MR, Golluscio RA, León RJC (1992) La patagonia extraandina: análisis de la estructura y funcionamiento de la vegetación a distintas escalas. Ecología austral 2:123–136

Plusagua (2011) Estudio hidrogeológico para abastecimiento de agua industrial al Yacimiento Cañadón Seco YPF SA. http://www.plusagua.com.ar

Schiuma, M, Hinterwimmer, G, Vergani G (eds) (2002) Rocas reservorio de las cuencas productivas de la Argentina. V Congreso de Exploración y Desarrollo de Hidrocarburos. Instituto Argentino de Petróleo y Gas (IAPG). Mar del Plata

Sylwan C, Droeven C, Iñigo J, Mussel F, Padva D (2011) Cuenca del Golfo San Jorge. VIII Congreso de Exploración y Desarrollo de Hidrocarburos. Simposio Cuencas Argentinas: visión actual. Instituto Argentino de Petróleo y Gas (IAPG), Mar del Plata, pp 139–183

Chapter 5
Conceptual Model

Abstract On the basis of the above-mentioned data, a conceptual model of the geohydrological system is proposed. First, the physical boundaries are defined: the upper one is the topography (permeable; positive sign); the lower one, the volcano-sedimentary complex (impermeable); the western, the Andes (permeable; positive); the eastern, the Ocean (permeable; negative); and the northern and southern ones, the respective watersheds, assumed to be permeable and positive. The input are the net vertical input (59–76 mm/year), entering by special mechanisms; groundwater inflow; and the contribution of losing streams, e.g. the Senguerr River. The main output is the ocean, followed by the domestic supply and the oil industry, though currently production water is re-injected. Intra-system transport is not very significant and occurs in the SGS, due to the occurrence of semi-confined aquifers. In this chapter, aquifer vulnerability is discussed; in order to assess it with the available data, the GOD method—developed by Foster and Hirata (1991)—was applied. It uses the indicators *groundwater occurrence, overall lithology of aquiperm or aquitard* and *depth to groundwater table*, with values ranging from 0 (negligible) to 1 (extreme). In this book, the vulnerability of the phreatic, semi-confined and confined aquifers of the SGS is assessed. In the phreatic aquifers, the values obtained are 0.52–0.7 (high); in the semi-confined, 0.273–0.343 (low to moderate); and in the confined, 0.065–0.084 (negligible). As the method considers an exogenous contaminant burden, reference is made to possible endogenous impact sources.

Keywords Boundary conditions · Limits · Sign · Input · Output · Intra-system transport · Vulnerability · GOD method

On the basis of the above-mentioned data, a conceptual model of behaviour of the geohydrological system is proposed. It has been considered advisable, once the model is described and analyzed, to discuss the issue of aquifer vulnerability, as it affects the entire system, even if it refers in particular to the SGS.

© The Author(s) 2017
M.A. Hernández et al., *Hydrogeology of a Large Oil-and-Gas Basin in Central Patagonia*, SpringerBriefs in Latin American Studies, DOI 10.1007/978-3-319-52328-6_5

5.1 Model Outline

The model is briefly presented, defining in the first place the physical boundaries of the system, which are subsequently characterized according to their condition (permeable or impermeable) and sign (positive or negative). The input, output, and transport values within the system itself are analyzed and, in the case of non-perennial streams, the variations in surface and groundwater storage.

The *boundaries* assigned to the system are as follows: the *upper* boundary is the topographic surface within the context of the basin; the *lower* one is the volcano-sedimentary complex (Middle and Upper Jurassic); the *western*, the Andes; the *eastern*, the Atlantic Ocean; and the *northern* and *southern* boundaries, the basin watersheds (Fig. 3.1); each one of them with a condition and sign that is characterized below.

The upper boundary is regarded as permeable with a positive sign, and it corresponds to one of the input functions of the system, the net vertical input. The lower boundary is assumed to be impermeable (aquifuge), as no continuity in the fissures could be recognized.

The western boundary is probably permeable with a positive sign, as suggested by the direction of flow shown in the equipotential diagram (Fig. 4.7) and by the favourable conditions for the occurrence of rapid concentration in the peri-Andean sector. Both the northern and southern boundaries could be assumed to be of the same type as the one above, although with certain reservations that are discussed below. Finally, the eastern boundary, which is permeable with a negative sign, was set at the ocean coastline, as it was impossible to characterize the offshore hydrogeological limit of the basin.

As regard the *input*, the net vertical input may be estimated on the basis of the water balance, which takes into consideration an analogous series approach and classified rainfall events (Sect. 4.3), with an approximate value of 59–76 mm/year, although there are doubts about the numerical value that may be assigned to the input areas in order to calculate volumes. This input involves rapid infiltration and delayed recharge, apart from the regional phenomenon of reduction in consumptive losses described in Chap. 4 (Hernández et al. 2009a).

Another source of input is the groundwater inflow, which—as shown in Fig. 4.7— in general originates from the recharge occurring in the western sector and, probably, from the north and south (Somuncurá and Deseado massifs); the data available is insufficient to analyze and quantify it. There is also the contribution of the Senguerr River, a losing stream, and secondarily of the Chubut and Deseado rivers, whose direct influence is limited to its environment, but which is quantitatively significant (Alvarez et al. 2006).

The main *output* is the discharge into the Atlantic Ocean, as in this region—as discussed above—consumptive loss is absent or insignificant, except in the case of direct evaporation from lakes and *mallines*. Water discharged from springs also falls within this category.

A significant output is the anthropogenic one, both from the SGI and the SGS. In the first case, due to the fact that it uses formation water for secondary recovery, but which—in keeping with the regulations—is partly re-injected into the oil deposits. As regards the SGS, due to legal restrictions, water for domestic use supplies the oil industry and for public supply, especially in the eastern sector. Agricultural use is very limited, restricted to the Sarmiento Valley on the Senguerr River and a very small suburban market garden in Comodoro Rivadavia.

Only in the SGS does intra-system transport occur, at the expense of the occurrence of semi-confined aquifer units (Patagonian and Santa Cruz aquifers), which also often behave as phreatic aquifers in certain locations and even as a confined aquifer in the case of the Patagonian (Hernández and Hernández 2013). In certain instances, transport is induced by extraction from the phreatic unit or, more frequently, from the semi-confined one. A different situation occurs regarding the injection of formation water into the wells of the SGS due to problems concerning the design, construction, or integrity of the pipelines, which constitutes a contamination problem.

Considering that the system has a non-permanent regime, the variations in water storage also have to be taken into account. Surface storage is very isolated and it is limited to the Musters and Colhué Huapi lakes, as well as to some endorheic depressions disseminated throughout the basin, mainly active during the spring and the early summer. Groundwater storage occurs in the SGS, where the net vertical input from the meteoric contributions has an influence. Following the reasoning explained in Chap. 4, important variations in the volume contained in the system are not expected.

5.2 Vulnerability

The terms "aquifer vulnerability" or "aquifer system vulnerability" are used in this book in the sense of "[representing] the sensitivity to be adversely affected by a contaminant burden" (Foster and Hirata 1991), a concept that was then expanded by Vrba and Zaporozec (1994) as "an intrinsic property of the groundwater system that depends on the sensitivity of that system to human and/or natural impacts."

These authors introduce three central elements for a deeper understanding of the concept. First of all, the *intrinsic* character, which makes vulnerability independent from the contaminant burden; when both are connected, the term used is "pollution risk" (Foster and Hirata 1988, 1991). The second element is the reference to an *aquifer system*, such as the case of this basin, in which there is a system with different aquifers. The third variable is the possibility of *natural impacts*, besides the anthropogenic ones.

The most widely used methods assume in all cases that there occurs an exogenous input through the vadose zone, such as the GOD (Foster and Hirata 1991), which is applied in this work. However, given the specificity of the pollution brought about by the activities of the oil industry, the endogenous source of the impact—that is, through the pipelines—is taken into consideration.

With a more accurate resolution and also very widely used, the DRASTIC method, developed by Aller et al. (1985), uses seven indicators, instead of three like the previous one. However, in many cases they are not available, especially in large territories with low population density, as is in general the case of the Extra-Andean Patagonia, and in particular of the San Jorge Gulf Basin.

GOD stands for *groundwater occurrence*, *overall lithology of aquiperm or aquitard*, and *depth to groundwater table* (in unconfined aquifers, or to strike in confined ones), the three parameters of the method. The numerical results, obtained as the product of the values assigned to the above-mentioned factors to qualify vulnerability, varies between 0–0.1 (negligible), 0.1–0.3 (low), 0.3–0.5 (moderate), 0.5–0.7 (high) and 0.7–1.0 (extreme) (Fig. 5.1). These vulnerability classes may be expanded, taking into consideration the trend; for instance, a value of 0.35 would be low to moderate vulnerability.

In the case of this basin, the GOD method was applied to the useful aquifers of the SGS (Patagonia and Santa Cruz formations), which, depending on their spatial position, may be phreatic, semi-confined or even confined, as explained above.

Thus, for the phreatic type of behaviour, a vulnerability index of 0.52–0.7 (high) was obtained. When the behaviour was of the semi-confined type, the result was 0.273–0.343 (low to moderate) and, in the case of the confined type, it was 0.065–0.084 (practically negligible).

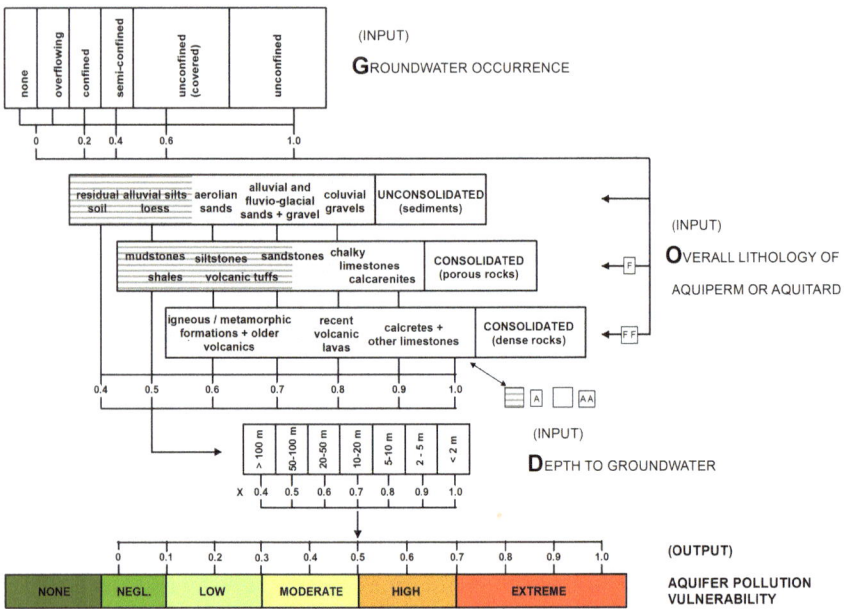

Fig. 5.1 GOD diagram for calculating vulnerability of aquifer systems (sensu Foster and Hirata 1988)

It is necessary to clarify that, according to this method, these differences arise from the "D" value in GOD, as it is the depth to the groundwater table in the case of phreatic aquifers, and the depth to the floor of the aquitard in the semi-confined or confined aquifers. It may be concluded that such an indicator has a key role in the GOD vulnerability index.

It is noteworthy that, in the case of phreatic aquifers, they require special attention for their protection in HC production areas. This is even the case with semi-confined aquifers, whose protection is discussed in Chap. 6.

The vulnerability classification was not applied to the other aquifers of the SGS or to the aquifers of the SGI because, on the one hand, they cannot be used for human supply and, on the other, they are confined and highly disconnected from the rest of the units in the system.

As regard the endogenous impact, it is represented by the integrity of the casings and the efficiency of their cementations in those sections of the pipeline that coincide with the presence of aquifers, which, in this case, are the Patagonia and Santa Cruz formations (Hernández and Hernández 2013). At present, impact assessment techniques are being used; they are based on integrity tests, which are applied to each well of interest to the oil industry (for production or injection), and on the detection of the presence of production water, HCs and/or metals in the aquifer system. As these techniques are developed by oil companies, they cannot be reproduced due to confidentiality reasons (Hernández et al. 2009b).

In the case of production waters, their salinity is usually high to very high and, should they gain access to the aquifers due to defects in the pipelines, they may even render important volumes of freshwater unusable.

References

Aller L, Truman B, Lehr JH, Petty RJ (1985) DRASTIC-A Standardized System for evaluating groundwater pollution potential using hydrogeologic settings. U.S. Environmental Protection Agency, Robert S. Kerr Environmental Research Laboratory, Office of Research and Development, EPA/600/2-85/018

Álvarez MP, Hernández L, Hernández MA, González N (2006) Relación aguas subterráneas-aguas superficiales en Patagonia Extrandina. República Argentina. Revista Latino-Americana de Hidrogeología 6:43–48 (Montevideo, Uruguay)

Foster S, Hirata R (1988) Determinación del Riesgo de contaminación de aguas subterráneas. Centro Panamericano de Ingeniería Sanitaria y Ciencias del Ambiente (CEPIS), Lima, Perú, 79 pp

Foster S, Hirata R (1991) Determinación del Riesgo de contaminación de aguas subterráneas. Una metodología basada en datos existentes. CEPIS, Lima, Perú, 81 pp

Hernández L, Hernández MA (2013) Características hidrolitológicas de las formaciones Patagonia y Santa Cruz. Cuenca del Golfo San Jorge. (Provincias de Chubut y Santa Cruz). In: González N, Kruse EE, Trovatto MM, Laurencena P (eds) Agua subterránea recurso estratégico I. EDULP, La Plata, pp 112–117

Hernández MA, González N, Hernández L (2009a) Regiones áridas. Procesos diferenciales de recarga y casos ejemplo de Argentina. In: Carrica J, Hernández MA, Mariño E (eds) Recarga de acuíferos. Aspectos generales y particularidades en regiones áridas. AIHGA-Amerindia, Santa Rosa, pp 63–70

Hernández MA, Scatizza C, Rojo M, Preiato S, Hernández L (2009b) Un método para estimar la sensibilidad hidrológica aplicado en la cuenca del Golfo San Jorge. Provincias de Chubut y Santa Cruz. Boletín Geológico y Minero de España 120(4):523–532 (Madrid)

Vrba J, Zaporozec A (1994) Guidebook on mapping groundwater vulnerability. International Association of Hydrogeologists (IAH) 16, Verlag Heinz Heise, Hannover, 131 pp

Chapter 6
Hydrogeology in Oil and Gas Production

Abstract The evolution of HC production since 1907, the economic transformation of the basin, and the increase in water demand due to the exploitation of new oil deposits and the use of secondary recovery are discussed. The conventional methods have produced a cumulative total of 734,508,000 m^3 of oil and 101,567,000,000 m^3 of gas, with 39,300 active wells in 2014. Today, secondary recovery with water injection amounts to 48% of the total extraction, initially using fresh groundwater (Patagoniano aquifer) and, at present, lower quality water (from the aquifers of the SGI) or from the underflow of the Senguerr River, as well as the reuse of production water. As regard water for domestic supply, the aqueduct from Musters Lake is used; at a provincial level, the extraction from aquifers fit for human consumption was regulated. Considering the prospects of the recently undertaken unconventional HC exploitation (tight and shale gas/oil in the Pozo D-129 Formation and, in the future, the Pozo Anticlinal Aguada Bandera and Pozo Cerro Guadal formations), a significant demand can be foreseen, of the order of 7500–30,000 m^3/well in the case of fracking and 2250–9000 m^3/well for drilling. Given the legal restrictions imposed for the aquifers of the SGS, a comprehensive management of the basin must be implemented—which is discussed in Chap. 8—including the aquifers of the SGI, the underflow of the Senguerr River, and the recovery of the flowback, in order to guarantee the sustainability of the exploitation and the continuity of the operations.

Keywords Oil and gas production · Secondary recovery · Current water supply · Water for hydrocarbon production and domestic use · Unconventional hydrocarbon exploitation · Future water demand

The discovery of oil in 1907 brought about a profound change in the socio-economic context of the basin, whose possibilities for development seemed to be limited to sheep farming and incipient agriculture in the Sarmiento Valley. In fact, the search for groundwater that led to such a discovery was aimed at building a port for the shipping of wool and agricultural produce. Comodoro Rivadavia and Sarmiento

© The Author(s) 2017   71
M.A. Hernández et al., *Hydrogeology of a Large Oil-and-Gas Basin in Central Patagonia*, SpringerBriefs in Latin American Studies, DOI 10.1007/978-3-319-52328-6_6

were the only urban settlements at the time, and the area was the only centre of oil production in the country until 1918, when oil was found in the Neuquén Basin.

A burst of activity and an increase in population settlement occurred, with a large influx of immigrants from Europe—coinciding with both world wars and the interval between them—from the north of Argentina and from Chile (Gadano 2012). In addition to Comodoro Rivadavia, an important conurbation and new urban centres—such as Caleta Olivia, Las Heras and Pico Truncado—have emerged, with industrial ventures, new private and public services, educational growth at all levels, and significant cultural advancement. In summary, a profound social and economic transformation has taken place.

In addition to the original hydrocarbon-bearing reservoir, the Salamanca Formation—which lies at a depth of only 545 m—the Cretaceous deposits of the Chubut Group have been found to occur at greater depths and a progressively larger geographic extent, from Campamento Central in Comodoro Rivadavia in the Northern Flank of the Eastern Sector through the Southern Flank to the Western Sector. Offshore exploration has also been undertaken (Homovc and Lucero 2002).

As described below, the oil industry is highly water demanding in all its stages (exploration, extraction, transportation, processing, and marketing) and, due to the above-mentioned arid climate conditions, this has always been a critical problem in the basin. Therefore, the importance of the understanding of the groundwater resources is discussed, both for conventional and unconventional oil and gas sources, in each case associated with their specific mode of production, requirements, and availability.

6.1 Exploitation of Conventional Oil and Gas

The contents of the paragraph above correspond to the conventional oil and gas exploitation methods, which have been in practice so far. The San Jorge Gulf Basin has historically been the most oil-productive basin in the country, with a cumulative total for the period 1907–2015 of 734,508,000 m^3, 40.9% of the total, followed by the Neuquén Basin. As regards natural gas, with a cumulative total of 101,567,000,000 m^3 between 1986 and 2015, it supplies only 8.8% of the country's total, only exceeding the total of the Austral Basin (IAPG 2016). The oil is waxy (15°–30° API) and, since 1907, 39,300 wells have been drilled, with 13,533 wells still active in 2014 (Hirschfeldt 2014).

In the early years, while the primary recovery stage lasted, the water needs were met with difficulty, using spring water from the above-mentioned Manantiales Behr (Sect. 4.1) and/or the SGS aquifers (Patagonian and Santa Cruz), and occasionally from others of the SGI with brackish or saline water. Due to the ageing of the oil fields, secondary recovery techniques by water injection started being implemented in the 70s; by 2014, 48% of the total production was obtained by means of such a method, as opposed to 51% by primary recovery and 1% by assisted recovery (IAPG 2016).

In order to satisfy the requirements of the water injection and due to the regulations in force in Santa Cruz—which will be discussed below—for 10 years and until 2016, the oil industry used a discharge of the order of 8000–10,000 m^3/d of water from the underflow of the Senguerr River. It was interrupted due to the decrease in the oil/water ratio in the oil fields, as a result of their ageing. At the moment, 92.5% of the production of fluids is water (Hirschfeldt 2014). Secondary recovery makes use of many of the techniques applied to shale gas/oil exploitation, especially fracking, to facilitate injection, the addition of biocides, scale inhibitors, pH regulators, corrosion inhibitors and other chemicals.

The current availability is in equilibrium with the requirements of the industry when the aqueduct that supplies Comodoro Rivadavia, Caleta Olivia and Rada Tilly is functioning normally, which is not always the case. Such equilibrium is also attained because brackish water from aquifers—mainly from the SGI—is being used and the increasing excess of production water is being re-injected.

The most important regulation is the Ley 2185 (Act 2185) of the Province of Santa Cruz, which declares the area comprised between the tablelands Meseta Espinosa and El Cordón to be a hydrogeological reserve. It is bounded by the 300 m a.s.l. contour line to the north, the axis of the Deseado River to the south, the 67° 30′ meridian to the east, and the 69° meridian to the west. It establishes that:

Oil companies, public or private, shall respect the technical conditions stipulated by the enforcement authority for the preservation of the hydrogeological resources, in particular regarding the current or future drilling of wells, as well as those concerning abandoned wells.

The related Disposición 135 (Order 135) states:

Whereas many of the existing wells do not adequately protect the existing freshwater aquifers by means of a surface pipe,

Whereas this situation poses a potential risk of contamination for the existing aquifers of interest, such as the case of the San Jorge Gulf Basin (Patagonia Formation).

And it clarifies that:

In all of these cases, an additional barrier shall be implemented over the freshwater-bearing aquifer levels (aquifers of interest: those fit for human consumption, livestock watering, and irrigation) of up to 2000 ppm of total dissolved salts or 3000 μS/cm of electrical conductivity […]

It also legislates on the integrity of the casings and the insulation of the pipeline into the aquifers of interest (Hernández and Hernández 2013).

In turn, the Province of Chubut has the Ley 1850 (Act 1850), which regulates the water policies of the province and is very general, the Código de Aguas (Water Code), and the Código Ambiental (Environmental Code), as well as a myriad of regulations of different legal standing (for instance, the Decreto 10/95 [Decree 10/95] on environmental protection in oil activities and the Decreto 1151/15 [Decree 1151/15], a protocol for environmental incidents in oil activities), with several projects also including unconventional oil and gas exploitation.

6.2   Exploitation of Unconventional Oil and Gas

In the last few decades, Argentina has been facing an energy shortage that is very difficult to balance in the medium term by means of an increase in the production of nuclear and hydroelectric power, of renewable resources, coal—which is practically irrelevant—oil, and gas (Riavitz 2015).

The results of unconventional oil and gas exploitation in the USA, which were exponentially positive in the short term, started a worldwide trend in that direction, including Argentina, where in the last decade the reserves, and oil and gas production decreased, despite the increase in active wells and in the water/oil ratio. Unconventional sources of *tight* and *shale gas/oil* have been proven to exist here, combined with a profound geological knowledge—as these source rocks were drilled and tested during conventional oil and gas exploitation—and with generally available installed infrastructure (Caligari and Hirschfeldt 2015). After the Neuquén Basin, which is the most favourable basin and currently in production, the San Jorge Gulf Basin has very good prospects of success in *shale gas/oil* exploitation (Hernández et al. 2014).

The source formations are Pozo D-129, the main source rock in the basin (Stinco 2015)—within which it is widely distributed—Pozo Anticlinal Aguada Bandera, and Pozo Cerro Guadal, the last two recognized as a whole as Neocomian and as being the oldest in the region. Unlike the case of the Neuquén Basin, exploitation is incipient.

The most favourable general conditions for the future, besides the lithology and thicknesses, are the good information available, collected from the historical exploration in the area, and the presence of certain basic infrastructure (e.g. access roads, batteries, pipelines, electric power). Among the negative conditions are the limited availability of water and the restrictive use regulations (Hernández and Hernández 2016).

The water requirements are greater than those of conventional oil and gas production, of the order of $7500-30{,}000 \text{ m}^3$ per well for fracking, which would constitute 70% of the water total, with the remaining 30% being used in the drilling. There are cases in which up to $174{,}000 \text{ m}^3$ were required for a six-well platform (Scatizza et al. 2013). Even though the amounts are not excessive, they are significant demands for an arid region.

The availability of water resources is considerably lower than in the Neuquén Basin, where surface sources (Neuquén River, and the Mari Menuco and Los Barreales dams)—which are amply sufficient for large-scale supply—can be used (Hernández 2015).

On the other hand, in the San Jorge Gulf Basin, the surface resources are practically limited to the Senguerr River—from whose underflow until not long ago water was collected—since the Deseado River is an intermittent stream and the Chico River has been cut off. The Musters and Colhué Huapi lakes, in turn, are highly compromised at present, due to the progressive decrease in contributions.

Consequently, groundwater is essential within the framework of comprehensive management. In the SGS, the Patagonian and Santa Cruz aquifers are bound by use restrictions, and in the Province of Chubut the guidelines may be issued by the enforcement authority. Water in the SGI aquifers is brackish.

Even though in unconventional oil and gas resources there are no compatibility problems because source rocks do not act as aquifers, the quality of water in the injection may be an issue due to reactivity; the details are not known for confidentiality reasons. The management proposed includes the brackish water of the SGI, part of the underflow of the Senguerr River, and water recovered from the flowback, which in the Neuquén Basin is of approximately 35% in volume. In general, percentages between 20 and 70% are mentioned (Hernández 2015), so the most convenient water mixture must be carried out in each case, taking into consideration the logistical factor regarding transportation and temporary storage.

References

Caligari R, Hirschfeldt M (2015) Condiciones para la explotación de recursos hidrocarburíferos no convencionales en la Argentina. In: Recursos hidrocarburíferos No Convencionales shale y el desarrollo energético de la Argentina. Caracterización, oportunidades, desafío. EUDEBA, Buenos Aires, vol 3, pp 213–306

Gadano N (2012) Historia del petróleo en Argentina. 2nd edn. Edhasa, Buenos Aires, 709 pp

Hernández MA (2015) Recursos hídricos. In: Recursos hidrocarburíferos No Convencionales shale y el desarrollo energético de la Argentina. Caracterización, oportunidades, desafío. EUDEBA, Buenos Aires, vol 4, pp 307–348

Hernández MA, Scatizza CF, Hernández L, González N (2014) Hidrogeología en la explotación de yacimientos no convencionales de HC's. Patagonia extrandina, Argentina. XII Congreso Latinoamericano XII de Hidrogeología y XXVI de Hidráulica. CD-ROM. Santiago de Chile

Hernández L, Hernández MA (2013) Características hidrolitológicas de las formaciones Patagonia y Santa Cruz. Cuenca del Golfo San Jorge. (Provincias de Chubut y Santa Cruz). In: González N, Kruse EE, Trovatto MM, Laurencena P (eds) Agua subterránea recurso estratégico I. EDULP, La Plata, pp 112–117

Hernández L, Hernández MA (2016) Agua subterránea en proyectos de yacimientos no convencionales de HC's. Cuenca del Golfo San Jorge (Chubut y Santa Cruz). In: García RF, Rocha Fasola V (eds) Hidrogeología: Minería, Cultura y Educación. Editorial Científica Universitaria, Catamarca, pp 100–107

Hirschfeldt M (2014) Análisis de madurez de la Cuenca del Golfo San Jorge: desafíos actuales y futuros para un desarrollo sostenido y sustentable. Jornadas de Producción del Instituto Argentino de Petróleo y Gas (IAPG), Comodoro Rivadavia, 19 pp

Homovc JF, Lucero M (2002) Cuenca del Golfo San Jorge: Marco geológico y reseña histórica de la actividad petrolera. In: Rocas reservorio de las cuencas productivas de la Argentina. Instituto Argentino de Petróleo y Gas (IAPG), Buenos Aires, pp 119–126

IAPG (2016) Producción de petróleo y gas natural acumulada a 2015 desde el inicio de la actividad. Instituto Argentino de Petróleo y Gas (IAPG). Sistema de información estadístico para Petróleo y Gas (SIPG), Unpublished, Buenos Aires

Riavitz L (2015) El futuro de la energía en la Argentina: alternativas. In: Recursos hidrocarburíferos No Convencionales shale y el desarrollo energético de la Argentina. Caracterización, oportunidades, desafío. EUDEBA, Buenos Aires, vol 1, pp 33–138

Scatizza CF, Hernández MA, Preiato S, Di Lorenzo C, Wocca M (2013) Aprovechamiento sustentable de los recursos hídricos en el desarrollo de los yacimientos de hidrocarburos no convencionales. In: González N, Kruse EE, Trovatto MM, Laurencena P (eds) Temas actuales de la Hidrología Subterránea. EDULP, La Plata, pp 297–302

Stinco L (2015) Los recursos hidrocarburíferos en la Argentina y las características de los reservorios no convencionales del tipo *shale*. In: Recursos hidrocarburíferos No Convencionales shale y el desarrollo energético de la Argentina. Caracterización, oportunidades, desafío. EUDEBA, Buenos Aires, pp 139–212

Chapter 7
Management and Governance

Abstract The major conflicts between water uses, mainly domestic and mining uses, are addressed, analyzing the surface and groundwater availability—as well as the reuse of water in the oil industry—and their harmonious use, taking into consideration the legal regulations in force. The surface water is limited to the Senguerr River (discharge: 48.6 m^3/s), since the Chubut and Deseado rivers are in a marginal location, and the environment of the great lakes is highly compromised. The groundwater sources available are the water of the SGS that is unfit for human consumption, all of the water of the SGI, the underflow of the Senguerr River, the reusable water from the flowback of unconventional HC exploitation, and the water remaining from conventional secondary recovery, favoured by the progressive decrease in oil/water ratio. The management of water is outlined and its governance is analyzed, focusing on the difficulties in its implementation, given the great variety of social, sectoral and institutional actors. Besides, the situation is complicated further by the overlapping national, provincial and municipal jurisdictions. Added to this, there is a lack of specialized knowledge on the subsurface hydrology derived from scientific and technical research, the mobility imposed by the market on the corporations and trade unions, and a wide range of resource administrators. All of this is analyzed in this chapter, taking into account the necessary dynamics that would guarantee the sustainability in different scenarios.

Keywords Conflicting water uses · Water availability · Integrated management · Surface water, groundwater and recovered water · Harmonious use · Water governance · Complexity and difficulties

The marked water deficit in the region is in itself a condition that generates or fosters conflicts between water-using activities. Some are absent or have very little impact, such as agricultural activities, which are limited to the valley of the Senguerr River in the city of Sarmiento or small vegetable farms in the vicinity of Comodoro Rivadavia (Cañadón Ferrays, El Trébol and Manantiales Behr).

© The Author(s) 2017   77
M.A. Hernández et al., *Hydrogeology of a Large Oil-and-Gas Basin in Central Patagonia*, SpringerBriefs in Latin American Studies,
DOI 10.1007/978-3-319-52328-6_7

Industrial use is limited, generally associated with the oil industry in the production of tools, accessories, or spare parts, or in the maintenance of the facilities. It is also used in small-scale manufactured goods for the provision of inputs aimed at urban or commercial services.

Therefore, the main competitive activities that remain are the domestic supply and HC production, which have been in the longest-standing historical conflict from both a quantitative and qualitative standpoint. Even though at first the oil industry resorted to the water supply aimed at human consumption in Comodoro Rivadavia, which is of subsurface origin, in time it was necessary to import increasing volumes of river water, leading to the present-day crisis. The government of the Province of Santa Cruz explicitly imposed restrictions on the use of water from aquifers of the SGS, which are mentioned in Sects. 6.1 and 6.2 (Hernández and Hernández 2013).

Any attempt at making a proposal that may somehow help control or mitigate the existing conflicts and timely prevent new ones should undoubtedly be based on a harmonious use including various sources with the lowest possible degree of impact.

Consequently, the surface and subsurface availability, as well as water reuse in the oil industry, are analyzed, taking into account the extensive experience, both positive and negative, gained so far in the basin.

Surface water, as has been repeatedly mentioned above, is scarce and also relatively distant from the water-using sites, both urban and oil-related. The Senguerr River, with a discharge of 48.6 m^3/s, is used for water supply to Comodoro Rivadavia, Caleta Olivia and Rada Tilly by an aqueduct (115 m^3/d) with serious difficulties as regards its construction, maintenance and management. Such discharge is also used for water irrigation in the Sarmiento Valley and, above all, it is responsible for maintaining the level of the Musters and Colhué Huapi lakes, the latter of which is clearly receding, which implies a considerable ecological impact; a reduction in their regime has also occurred due to the misuse of the resources of the Senguerr River tributaries upstream.

The Chubut and Deseado rivers occur in marginal locations to the north and south; the former occurring very far from the oil production sites. The Deseado River, despite having an intermittent regime, is used as a subsidiary source at a small scale.

Regarding *groundwater*, it is necessary to foresee that the current legal limitations for the aquifers of the SGS in better conditions (Patagonia and Santa Cruz formations) could be extended in the face of growing social pressure, which is why no expectations should be generated concerning that source, at least in locations in which good quality water is available.

As expressed in Sect. 4.4, part of the remaining aquifers of the SGS and all of those of the SGI have brackish or saline water. However, it is still possible for such water to be used in the practice of fracking, provided it does not react with the additives commonly used, though such information is not available or at least not explained in detail at present. Nevertheless, at the moment, its application is undergoing experimentation in the basin (Yacimiento El Trébol, personal communication) with positive results.

Due to the progressive decrease in oil/water ratio registered at present, the use of volumes of water that will not be used in the exploitation of conventional oil and gas with secondary recovery could also be taken into consideration. Until September 2016, such volumes were obtained from the underflow of the Senguerr River and will be made available in the future. New contributions from this source could also be used, as long as they do not have a negative impact.

Therefore, the management of water in the exploitation of unconventional oil and gas in the basin, which today is of paramount importance, should be based on the harmonious use of the above-mentioned sources (Table 7.1).

The conjunctive use proposed above would also have seasonality: for instance, a greater subsurface contribution to the supply in the summer, when the flow of the Deseado River is not continuous, or in the peak of winter, when the Senguerr River freezes.

Logically, it should be a harmonious use, which would ensure the sustainability of the water resources in general and of the exploitation for productive purposes in particular. Given the multiplicity of factors involved, it will be necessary to set out the processes of coordination and cooperation of the different social, sectoral and institutional actors that should participate in the integrated management of groundwater. In such processes, the territory and the basin should be considered as active entities, in order to guarantee the water supply and the environmental services. This is precisely what the concept of water governance entails (OECD 2015).

Even though these postulates are clear and logical, and that this is certainly a valid proposal for water management in the San Jorge Gulf Basin, it is actually difficult to implement it (Custodio 2015). In this case, the complexity arises from the quantity and diversity of actors with different interests, backgrounds and jurisdictions. But, above all, it stems from the lack of a more precise knowledge of the subsurface hydrology derived from scientific and technical research, since the data available are obtained from the geological mining activity (HCs and metal minerals), which despite the regulatory developments are still unable to make up for

Table 7.1 Water management. Water offer composition for the exploitation of unconventional oil and gas

Water for unconventional oil and gas exploitation	Groundwater	Surface water	Recycling of flowback	Surplus of secondary recovery
	SGS aquifers	Senguerr River	With pretreatment	Unused water
	SGI aquifers (brackish–saline water)	Deseado River (limited)		
	Underflow water			

the lack of detailed geohydrological data. Contributions mostly originate in the scientific system, which by the way has the least financial resources to support any research.

Government actors come from the domain of both provinces (each one with its police power over water and enforcement authority for regulations), whereas certain aspects are within the national jurisdiction, which the university and research centres also belong to. The body politic is another necessary actor, at municipal, provincial and national levels, as regulations and controls are established by the three of them.

Another undeniable participant is the oil industry, as the major water user in the region and at the same time a potential source of impact on the resources. Closely connected to such an activity are the trade unions, not only the one directly concerned, but also all of the other related trade associations.

The most relevant group out of all the social actors is naturally the population, which—as mentioned in Chap. 3—was estimated to be 380,000 inhabitants in 2015 (310,000 registered in the 2010 census). The provision of the public service is the responsibility of different administrators (municipal, cooperative, and private). Other minor actors are the intensive farmers, more common in the Sarmiento Valley, the communities in the Andean foothills, and in the vicinity of Comodoro Rivadavia and Caleta Olivia.

The above-mentioned complexity also lies in the diversity of actors, their mobility, and the shifting economic setting. The oil industry is highly dependent on the market, whereas the political and government communities in general have a timing of their own, which—together with the corporate mindset—conspires against governance when state policies are lacking.

The fact that the state of the groundwater resources in the region is not seriously compromised at the moment is no excuse to avoid addressing the need for effective governance; on the contrary, it might be the perfect opportunity to do so, before the current conflicts escalate and/or new ones arise (González and Hernández 2016).

But, above all else, in order to reconcile the interests of the actors around this common purpose and to ensure the continuity with timing different from the one of politics, it is essential to give the governance the necessary dynamism (Custodio 2015). This is important because the proposal often seems to resemble more a procedures manual or best practice guide than a real undertaking that must stay "alive" and adapt to the permanent changes in the conditions of the context.

References

Custodio E (2015) La difícil gobernanza de las aguas subterráneas. Seminario Nacional del Observatorio del Agua. Fundación Botín, Madrid, 19 pp
González N, Hernández MA (2016) La gobernanza del agua en sistemas agro-productivos de Argentina. Aguas subterráneas. In: Custodio E, Varni M, García RF (eds) Seminario Gobernanza del agua en áreas con escasez: Gestión de las aguas subterránea. Editorial Científica Universitaria, Catamarca, pp 12–18

Hernández L, Hernández MA (2013) Características hidrolitológicas de las formaciones Patagonia y Santa Cruz. Cuenca del Golfo San Jorge (Provincias de Chubut y Santa Cruz). In: González N, Kruse EE, Trovatto MM, Laurencena P (eds) Agua subterránea recurso estratégico I: 112–117. EDULP, La Plata

OECD (2015) The governance of water regulators, OECD studies on water. OECD Publishing, 116 pp. http://dx.doi.org/10.1787/9789264231092-en

Chapter 8
Concluding Remarks

Abstract The regional geohydrological characteristics of a vast oil-and-gas-producing basin are presented for the first time. It is located in the Extra-Andean Patagonian region, and it is characterised by its arid climate, with a marked water deficit, a limited occurrence of watercourses and low population density. The physical context (geology, geomorphology and soils) and the hydrological elements (meteorology, and surface and subsurface hydrology) are described in order to define a conceptual model of the geohydrological system. Its vulnerability is analysed and a management plan is proposed for the water resources and their governance, taking into consideration the great socioeconomic interest of the unconventional exploitation of HCs (shale gas/oil), which has been undertaken recently and offers excellent prospects.

Keywords San Jorge Gulf Basin · Groundwater · Regional study · Socioeconomic interest · Unconventional HC exploitation

(a) The decision to write this book arose from an attempt to contribute to an understanding of the groundwater in a vast arid territory of 740,000 km². No integral studies on its water resources have been conducted, despite the large water deficit and the fact that it is a mineral- and oil-rich area. The San Jorge Gulf Basin is located in this region, the Extra-Andean Patagonia, between parallels 43° and 47° S, at the edge of two large cratons and flanked by two enormous limits, the Atlantic Ocean and the Andes.

(b) Extending over an area of 59,510 km² (40,530 km² onshore), the basin has an arid, mesothermal climate, with a rainfall of 227 mm/year and a potential evapotranspiration of 704 mm/year (90-year record), which shows a significant annual water deficit of the order of 427 mm from August to April. Snowfalls occur, heavier in the Andean foothill area, as well as frosts. The anti-trade winds (Westerlies) have a frequency of 473/1000 and create a regional rain shadow in the Extra-Andean Patagonia.

© The Author(s) 2017   83
M.A. Hernández et al., *Hydrogeology of a Large Oil-and-Gas Basin in Central Patagonia*, SpringerBriefs in Latin American Studies, DOI 10.1007/978-3-319-52328-6_8

(c) The regional sedimentary episode that included this basin corresponds to a NNW–SSE depocentre that developed from the late Carboniferous–Permian to the Permo–Triassic, which facilitated the granite intrusion of the cratons. During the Middle Jurassic, an extensional process occurred: the depocentres were filled in a late rift stage and N–S-trending extensional faulting developed. In the Tertiary, the faulting was affected by compressive activity with effects of tectonic inversion, which is responsible for the present-day Bernardides fold belt. Towards the east, W–E-trending extensional faults occurred and they extended until the opening of the Atlantic Ocean. They were followed by a regional erosional event that affected the entire basin, generating the available space for a new Cretaceous sedimentary cycle, represented by the Chubut Group and hosting the main oil deposits.

(d) The landscape is marked by large positive landforms, such as the tablelands descending towards the east and the Bernardides fold belt, whereas the most outstanding negative ones are the depression of the central lakes, the valleys of the Senguerr, Chico, Chubut, and Deseado rivers and the endorheic depressions.

(e) Soils are typical of arid regions, fit for limited sheep grazing (classes V and VI), belonging to the Aridisol, Entisol and Mollisol orders, with the first ones predominating. Their hydrological role is given by generally coarse textures and low specific retention, facilitating the infiltration of the scarce meteoric contributions.

(f) The Senguerr River is perennial, with Andean headwaters, a pluvio-nival regime, an annual discharge of 48.6 m^3/s and it loses water to groundwater. The Chubut River has similar characteristics, but it only involves a small sector in the north of the basin, whereas the Chico River—which years ago connected both of them—is inactive at present. It was part of a system, together with the Musters and Colhué Huapi lakes; the latter is undergoing a distinct process of retreat and desiccation. The Deseado River is geographically marginal, occurring on the southern border, and has a transitory behaviour.

(g) From a geological standpoint, this intracratonic basin is developed over Precambrian to early Mesozoic igneous rocks, and upper Palaeozoic and Mesozoic sedimentary rocks. These underlie a lithological thickness of up to 8000 m in the centre of the basin that starts with the so-called Middle and Upper Jurassic Volcano-Sedimentary Complex (Lonco Trapial Group in the north and Bahía Laura Group in the south), composed of volcanic and volcaniclastic rocks, followed by Cretaceous and Tertiary sedimentary rocks.

(h) The Cretaceous, of continental origin, starts with Neocomian Group deposits that underlie those of the Chubut Group, which are of great importance due to their HC resources (sandstones, conglomerates, conglomerate sandstones, siltstones, tuffs, lutites and other lithologies).

(i) The Tertiary sequence starts with the Palaeocene marine sediments of the Salamanca Formation (first Atlantic marine ingression into the basin) with basaltic intercalations, followed by another continental cycle that extended up to the Patagonian ingression (late Oligocene–Miocene) and overlain by gravel and sandstone continental deposits.

(j) The geology of the basin was discretized into Western Sector, Bernardides, Eastern Sector and Offshore. The Eastern Sector is divided into Northern Flank, Basin Centre and Southern Flank.

(k) The geology summarized has its hydrolithological correlate in two systems composed of aquifer, aquitard and aquiclude levels over an aquifuge, referred to as Lower Geohydrological System (SGI) and Upper Geohydrological System (SGS).

(l) The SGI includes the deeper formations up to the Upper Cretaceous and it bears brackish to saline waters in confined aquifers. The SGS comprises the main aquifers in the region, with fresh to brackish water, occurring in the Patagonia and Santa Cruz formations (Patagonian aquifer) as semi-confined or phreatic levels.

(m) The dynamic component of the system develops in this physical context (recharge, circulation and discharge). Recharge to the system is direct autochthonous, indirect autochthonous, and allochthonous. The first type, despite the negative value of the edaphic water balance, occurs by means of special mechanisms that are typical of arid regions: reduction in consumptive losses due to the xerophytic vegetation, rapid concentration, rapid infiltration, delayed recharge and contribution of losing streams. It was estimated between 59 and 76 mm/year.

(n) Circulation in the Patagonian aquifer was reconstructed by means of an equipotential map, with the oil-producing area showing more information. The regional direction of flow is W–E and NW–SE, with divergent directions in certain areas, a radial pattern and an average hydraulic gradient of 1.4–1.6×10^{-2} (mean extremes 8.10^{-3} and 1.5×10^{-1}).

(o) The mean velocity of the groundwater flow is of the order of 2.55×10^{-2} to 2.25×10^{-1} m/d. Regional discharge occurs to the Atlantic Ocean, with a limited quantitative participation of the consumptive use due to the predominance of xerophytic vegetation.

(p) Hydrochemically, water in the aquifers of the SGI is saline (TDS 5000–20,000 mg/L) and predominantly a sodium chloride type. In the SGS, in the Patagonia Formation, freshwater (TDS 530–1000 mg/L) and brackish water (TDS 1000–3000 mg/L) occur, with the sodium chloride facies prevailing over the sodium sulphate and sodium bicarbonate ones. In the Santa Cruz Formation, freshwater (TDS 530–1130 mg/L) of the sodium bicarbonate and sodium chloride types can be observed.

(q) The original sheep farming economy was displaced by HC production, as a result of the discovery of oil in Comodoro Rivadavia in 1907. A cumulative total of 734,508,000 m^3 of oil and 101,567,000,000 m^3 of gas have been exploited. Out of the 39,300 wells historically built, 13,533 are still active.

(r) A large part of the extraction of conventional oil and gas in the basin has been obtained by secondary recovery since 1970, demanding a great quantity of groundwater. Today, such demand has decreased due to the increase in water/oil ratio and the reuse of the water resources.

(s) The worldwide expansion of the exploitation of unconventional oil and gas (tight and shale gas/oil) has reached Argentina, and in the San Jorge Gulf Basin such resources are available; initial extraction is already under way by means of fracking techniques.

(t) In the Pozo D-129 and Anticlinal Aguada Bandera formations, the source rocks that occur are suitable for unconventional techniques, which as a matter of fact demand large quantities of water. Given the legal restrictions regarding the use of freshwater aquifers for such purposes and for flowback re-injection, an adequate management of the water resources is proposed, in order to take advantage of the occurrence of significant technically recoverable reserves.

(u) The management plan includes the saline aquifers of the SGI, the aquifers bearing non-potable water of the SGS, the underflow of the Senguerr River, the partial recovery of the flowback within the guidelines, and the reuse of the water surplus of the conventional secondary recovery.

(v) For the plan to be effective, it is essential to obtain a more detailed knowledge of the geohydrology in order to guarantee both a sustainable supply and the necessary protection of the aquifers, and to implement a monitoring and control program, as well as a mathematical model with simulation and forecast capacity.

(w) This book proposes a conceptual model that would provide a basis for a numerical model, paying special attention to the boundary conditions.

Index

A

Arid climate, 6
Abandoned wells, 73
Accessories, 78
Active dune, 26
Active wells, 71
Actual evapotranspiration, 20
Actual reduction in consumptive losses, 49
Additives, 78
Aeolian, 3
Agricultural use, 67
Alicurá, 5
Alkaline plateau basalts, 22
Allochthonous, 41, 53
Allochthonous origin, 1
Alluvial plain, 27
Alluvial valley, 25
Alto Río Senguerr, 16
Analogous series, 66
Analytical results, 56
Andean headwaters, 53
Andes, 2
Anglo-Boer War, 42
Anionic behaviours, 57
Anionic dispersion, 60
Anisotropy (horizontal and vertical), 48
Anisotropy, 41
Annual, 15
Annual deficit, 15, 20
Annual rainfall, 2
Annual relative humidity, 18
Annual temperature, 15
Anthropogenic, 66
Anthropogenic discharge, 56
Anticlinal Grande area, 36
Anticlines, 23
Ap Iwan Hill, 26
Aptian–Albian, 34

Aquicludes, 43
Aquifer, 43, 46
Aquifer system, 20
Aquifer system vulnerability, 67
Aquifer vulnerability, 65
Aquifuges, 43
Aquitards, 43
Argentinian Extra-Andean Patagonia, 2
Argillic horizon, 29
Argixerolls, 29
Aridisol order, 29
Aridisols, 15
Aridity index, 20
Arid regions, 49
Arroyo Chalía, 28
Arroyo Genoa, 28
Arroyo Page, 27
Arroyo Shaman-Apeleg, 28
Arroyo Verde, 28
Artificial wetlands, 30
Assisted recovery, 72
Atlantic marine ingression, 36
Atlantic Ocean, 2, 29
Austral Basin, 72
Authorities, 46
Average hydraulic gradient, 41, 55
Axillary bud, 49
Azorella compacta, 49

B

Basin, 15
Badlands, 36
Bahía Laura Group, 31
Bajo Barreal Formation, 30, 32
Balshort, 10, 54
Barometric pressure, 10, 15, 19
Basaltic tablelands, 22
Basic intrusive, 23

© The Author(s) 2017
M.A. Hernández et al., *Hydrogeology of a Large Oil-and-Gas Basin in Central Patagonia*, SpringerBriefs in Latin American Studies, DOI 10.1007/978-3-319-52328-6